The Politics of Partnerships

A Critical Examination of Nonprofit–Business Partnerships

The Politics of Partnerships

A Critical Examination of Nonprofit–Business Partnerships

by
Maria May Seitanidi

 Springer

Dr. Maria May Seitanidi
Brunel Business School
Brunel University West London
Uxbridge
Middlesex UB8 3PH
United Kingdom
may.seitanidi@brunel.ac.uk

ISBN 978-90-481-8546-7   e-ISBN 978-90-481-8547-4
DOI 10.1007/978-90-481-8547-4
Springer Dordrecht Heidelberg London New York

Library of Congress Control Number: 2010920319

Printed on acid-free paper

Springer is part of Springer Science+Business Media (www.springer.com)

Dedicated to Georgia, Alex and Sophia:
my agape, noesis and kallos

Abstract

The widespread use of partnering in the US and the UK has attracted a significant amount of literature on the phenomenon and its strategic use. The book suggests that partnerships between nonprofit organisations and businesses allow both organisations to strategically meet their organisational goals. However, there has been little attempt to explicitly examine whether these relationships deliver benefits that result from the intentional combined efforts of the two partners and extend beyond the organisational to the societal level.

Based on 75 interviews in businesses and nonprofit organisations in the UK the book offers a critical examination of the partnership phenomenon under the three chronological stages of cross sector social partnerships: formation, implementation, outcomes. The book puts forward a holistic framework for the study of partnerships that allows for observations beyond any single stage. It addresses some critical knowledge gaps by including the perspective of both partners within a single study and comparing the organisational and societal outcomes of these relationships. Two in-depth partnership case studies are examined: an environmental nonprofit organisation, Earthwatch, in partnership with a mining company, Rio Tinto; and a youth charity, The Prince's Trust, in partnership with the Royal Bank of Scotland. The case study findings are compared and corroborated with the partnership experiences of 29 further organisations that participated in the research.

The study suggests that when collaborative nonprofit organisations partner with businesses there is an "overt functional conflict deficit" or in other words there are insufficient opportunities for the expression of divergent opinions that would lead to fundamental changes within organisations and society. As such conflict and change are connected within the context of nonprofit-business partnerships through the intention for change. Untying social partnership from the partnership approach based on the triple distinction among organisational, social, and societal outcomes contributes in bridging the gap between rhetoric and reality regarding the roles and responsibilities of the partners and partnerships.

.

Acknowledgements

I am grateful to Stuart Reid, the Associate Director of the Postgraduate Certificate in Cross-Sector Partnership (PCCP) at the Partnering Initiative at the University of Cambridge – Industry Programme, for putting me in touch with his students that allowed me to develop my first case study. I would like to thank all the organisations and the individuals who contributed to this research by allowing me to interview them. In particular, my sincere thanks to the case study organisations Rio Tinto, Earthwatch, Royal Bank of Scotland, and The Prince's Trust.

The organisations that participated in the research were chosen as exemplary cases that exercise advanced and forward-looking practices. The intent of the research was to challenge these practices in order to offer thought-provoking findings and suggestions for further improvement. Without their interest and support this research would not have been possible.

Andrew Crane and Dirk Matten offered me their invaluable guidance and support, for this project by never hesitating to challenge my thinking, allowing me to experiment with ideas while demonstrating intellectual trust that nurtured and helped my intellectual development. I am indebted to both of them. Also, sincere thanks to Jeremy Moon, at the ICCSR for his continuing support throughout this project and beyond.

I would like to thank a number of colleagues at different academic institutions who shaped my thinking through suggestions and insightful remarks: at the LSE Peter Abel, George Gaskell, Helmut Anheier, Michael Power, Bridget Hutter, Michael Barzelay, David Lewis, Tasneem Mowjee, Paola Grenier, Joanne O'Mahoney, Lynne Nicolytchuk, Juan Carlos Cortazar Velarde, Preecha Dechalert, Nuno Thelmudo and Sascha Grillo. Special thanks to Mark Boden for his insightful guidance, patience and support during the course of writing this book. At the ICCSR-University of Nottingham Judy Muthuri and Krista Bondy for their comments on my chapters. In addition I would like to thank a number of prominent academics who offered me comments, guidance and encouragement throughout the years of this project: Sandra Waddock, Edwin Stafford, Joe Galaskiewicz, Ans Kolk, Laura Spence, Martyn Pitt, John Burchell, Peter Lacy, Peter Walker and Evaggelia Papoutsaki. My sincere thanks to Sarah Holmes, Popi Hatzipanagiotou and Maria Galiatsatou for their invaluable assistance.

I am grateful to the editors at Springer Fabio deCastro and Fritz Schmuhl for their encouragement and interest in this project. Sincere thanks to Takeesha Moerland-Torpey for leading me through the publishing guidelines and answering diligently my endless questions.

Last and certainly not least I would like to express my gratitude to my family for proving me with continuous encouragement and inspiration.

Contents

List of Tables

List of Figures

Abbreviations

AGM	Annual General Meeting
BAA	British Airports Authority
BIE	Business in the Environment
BUS	Business
CBI	Confederation of British Industry
CEO	Chief Executive Officer
CEP	Civic Education Project
CERG	Corporate Environmental Responsibility Group
CR	Corporate Reputation
CRM	Cause-Related Marketing
CSES	Complex Evolving Systems
CSO	Civil Society Organisation
CSR	Corporate Social Responsibility
DNGO	Development NGO
EFP	Employee Fellowship Programme
EMO	Environmental Movement Organisation
ENGO	Environmental Non-governmental Organisation
EP	Equator Principles
EPI	Ethical Purchasing Index
EWE	Earthwatch Institute Europe
EWA	Earthwatch Institute Australia
EW	Earthwatch
F&F	Fauna and Flora
FoE	Friends of the Earth
GOV	Government
HR	Human Resources
ICNPO	International Classification of Nonprofit Organisations
IPRA	International Public Relations Association

| MNC | Multi National Corporation |
| MoU | Memorandum of Understanding |

NCVO	National Council of Voluntary Organisations
NGO	Non Governmental Organisation
NPO	Nonprofit Organisation

| OFC | Overt Functional Conflict |

PARTIZANS	People against RTZ
PFI	Private Finance Initiative
PPP	Public Private Partnerships
PT	The Prince's Trust

| RBSG | Royal Bank of Scotland Group |
| RT | Rio Tinto |

| TPP | Tri-Partite Partnership |

WBCSD	World Business Council for Sustainable Development
WWF	World-Wide Fund for Nature
WSSD	World Summit for Sustainable Development

Introduction

In the late 1990s the idea of cross-sector collaborations was relatively new in Europe. The term 'partnership' was employed primarily to refer to partnerships between government and businesses, usually termed PPP (Public Private Partnerships). On the other hand 'strategic alliances' was the term employed for business-to-business partnerships. Until then 'sponsorship' was the most practised associational form between nonprofit organisations (NPOs) and businesses (BUSs), which was included within the broad area of corporate community involvement.

The relations between NPOs and BUSs witnessed a gradual intensification over the last 200 years (Gray 1989; Young 1999; Austin 2000; Googins and Rochlin 2000) resulting in increased interactions within both the philanthropic and transactional types of relationships (Seitanidi and Ryan 2007). However, the more recent gradual prominence of the concept of corporate social responsibility (CSR) within all sectors of society elicited an intensification of the debate with regard to the responsibilities of each sector in addressing environmental and social issues. In effect, CSR contributed to the increase of the interactions across the sectors and propelled NPO-BUS Partnerships (a type of social partnership) as a key mechanism for corporations to delve into a process of engaging with NPOs in order to improve their business practices by contributing their resources to address social issues (Heap 1998; Mohiddin 1998; Fowler 2000; Googins and Rochlin 2000; Mancuso Brehm 2001; Drew 2003; Hemphill and Vonortas 2003).

The work of a number of authors (Warner and Sullivan 2004; Birch 2003; Elkington and Fennell 2000; Austin 2000; Andrioff 2000; Waddock 1988) marked a shift in the literature reflecting practices on the ground. The shift refers to the stirring of these relationships towards an instrumental approach that aims predominately to deliver positive outcomes for the two partners. Although the *social character of partnerships* has been discussed in the literature, it has not been examined in detail within partnership case studies. The instrumental approach in partnerships increases the value of the relationship for the partners but underplays the role of these collaborations for society. Previously researchers on NPO-BUS Partnerships (Waddock 1988; Austin 2000) proclaimed that partners should answer the question of "how society is better off because of their joining resources and efforts" (Austin 2000:88). In effect they were asking the practitioner community to address the above question in the course of a partnership relationship. As NPO-BUS Partnerships have been taking place in the UK for some time it is important to address this question on a theoretical and empirical level as it will allow us to identify some critical knowledge gaps. As a result it is important to understand how the partnership relationship evolves between the two partners over time, what motivates each partner, how partnerships are being implemented and what the outcomes of these relationships are, both on the organisational and societal level.

In order to address the above questions the book examines the relationship across two in-depth partnership case studies based in the UK, the first comprising an environmental nonprofit organisation, Earthwatch, and a mining company, Rio Tinto, and the second a youth charity, The Prince's Trust and a bank, the Royal Bank of Scotland. It focuses on the three stages of the partnership relationship – formation, implementation and outcomes – in order to critically examine each relationship and further corroborate the findings of the in-depth case studies with the partnership experiences of 29 organisations. The research examines the extent to which the positive outcomes delivered by a partnership relationship extend to the social domain and if they are indeed the result of combined efforts between the partners. More importantly it aims to identify whether the relationships under study can be classified as social partnerships based on their delivery of positive outcomes to society. Furthermore, it examines the types of positive outcomes that are generated through these relationships.

The study suggests that when collaborative nonprofit organisations partner with businesses there is less conflict than would be ideally expected – given the assumption that conflict over fundamental perspectives is a prerequisite for change, and the desire for change is the foundation on which social partnerships are formed. The concept of "overt functional conflict deficit" is put forward to refer to the insufficient opportunities for the expression of divergent opinions that would lead to fundamental changes. Untying social partnership from the partnership approach based on the triple distinction among organisational, social, and societal outcomes contributes in bridging the gap between rhetoric and reality regarding the roles and responsibilities of the partners and partnerships.

The importance of studying partnerships between NPOs and BUSs within a developed economy can provide an indication of the challenges the existing sectors are faced with today and how they perform within the context of partnership relationships. The UK model presents a unique institutional context for studying partnerships that stems from the fact that partnerships are institutionally encouraged by the government (Moon 2004). As such this book contributes new data to the literature of NPO-BUS Partnerships deriving from the unique case of the UK economy and its institutional environment.

The theoretical contributions the book aspires to make are to: (a) offer a robust conceptualisation of NPO-BUS social partnerships; (b) suggest a distinction between social and societal outcomes in partnerships; (c) extend the study of organisational change as an interaction field within the field of NPO-BUS Partnerships; (d) relate the concept of overt functional conflict with change within the sphere of NPO-BUS Partnerships.

The first chapter of the book introduces the phenomenon of cross-sector partnerships and defines the area of exploration this study focuses upon: partnerships between nonprofit organisations and businesses, a type of social partnership. It presents a historical/organisational discussion of collaborative interactions between the two sectors and locates partnerships as a form of association between nonprofit organisations (NPOs) and businesses (BUSs). It further suggests two sets of parameters – i.e. macro and meso (institutional) forces – that contributed to the phenomenon under study. The chapter maintains, based on previous definitions that social partnerships address issues beyond the traditional boundaries of organisations and lie within what was termed the 'public policy' arena (Waddock 1988).

The second chapter grounds NPO-BUS Partnerships within the relevant literature and is organised in three sections: formation, implementation and outcomes. It suggests a classification of the existing NPO-BUS partnership literature under six strands which assist in locating the present study. The next section of the chapter presents in each partnership stage the main research questions and constructs that the research addresses. The differentiation in role and responsibilities of the profit and nonprofit organisations is put into question and the role of partnership as an agent of change is discussed. The chapter concludes that the study of NPO-BUS Partnerships in the UK can shed light on the phenomenon and the implications for the participating sectors.

The next three chapters comprise the empirical chapters, organised into the three chronological stages of partnerships: formation, implementation, outcomes. The first empirical chapter discusses the two case studies concentrating on the formation of the partnership by offering an overview of the relationship, highlighting the organisational characteristics associated with each partner organisation and the historical evolution of each partnership relationship. It further discusses the motives of each organisation that resulted in the development of the partnership. The final section discusses the findings from the two in-depth case studies in light of the comparative interviews across 29 organisations that participated in the research. The chapter maintains that collaborative NPOs have historically synagonistic relationship with BUSs and discusses the organisational characteristics, the partnership evolution and the partners' motives in each of the two cases studies.

The next chapter concentrates on the implementation stage examining the phases of partnership relationships and the dynamics as they evolve within both cases. Partnership selection, design, institutionalisation and change (as a process) are the four phases that partnership can encompass. The chapter connects the formation stage by discussing the existence of change as a motive with the implementation stage and the possibility of change taking place as the next phase after institutionalisation. The final section discusses the findings of the cases in light of the 35 comparative interviews of the study.

Chapter 5 focuses on the outcomes that accrue as a result of the partnership relationship. Each case study is examined separately in order to identify the organisational outcomes for each participating organisation. A distinction is proposed between social and societal outcomes which allows for a separation between the partnership form and the partnership approach. The main findings of this chapter are discussed in light of the comparative interviews concluding that organisational genesis, can be a borderline organisational/societal outcome of partnerships even if intentionality is absent.

Chapter 6 summarises the main findings of the research and moves beyond the answers to the research questions that were addressed within each empirical chapter. This chapter aims to highlight the main theoretical themes that encompass the three chronological stages of partnerships. The first theme refers to the overt functional conflict (OFC) deficit that occurs in the course of a partnership relationship between collaborative NPOs and BUSs. It draws attention to the local conditions in each of the three stages of partnerships that provide evidence for the OFC deficit. The second theme refers to change both as process and outcome in partnerships. It confirms the previous conceptualisation of change as an interaction field (Lovelace et al. 2001) where "content and action are inseparable" (Pettigrew et al. 2001:697) within the context of NPO-BUS Partnerships. It highlights the multiple dimensions of change such as the intention (intentional/unintentional), the level (organisational, personal, societal), the type (content, processual) and the effects of change (genesis, kinesis). It suggests that the link between the OFC deficit and change is the existence of intention for change from both partners which in effect can result in organisational genesis. If the intention for change is evident only in one of the partners then the OFC deficit decreases the potential of organisational genesis. The final theme refers to the distinction between partnership as a form and as an approach. In the first case social partnership consists of a form of association based on the previous definitions suggested by the literature (Waddock 1988; Austin 2000) and the distinction put forward by the book based on the types of outcomes. In the second case, the partnership approach refers to the proximity of collaboration between organizations. The last section of the book concludes with a future scenario as an epilogue to the future of social partnerships.

Commentary on Methods

The research was developed between July 2002 and January 2004 comprising 37 interviews with nonprofit organisations, 38 with corporations of which four were with people working in consultancies specialising in the broad area of corporate social responsibility. The total of 75 interviews was complemented with further organisational documents such as annual and internal reports, and a participant observation session of one meeting. Appendix 1 presents all the internal documents consulted for the case studies.

The research followed an inductive design informed by the literature, which was gradually enriched as a result of the iterations between the data and the relevant literatures. The role of conflict in partnerships emerged as a result of several iterations of analysis as an overarching theme.

Case studies were developed in an effort to build rich descriptions based on in-depth understanding of the phenomenon of partnerships under investigation (Weick, 2007). The criteria employed to select the cases were: (1) the scope of

Table 1 Intro: criteria for selecting the cases

	Form	Scope of activity	Purpose	Resources	Reputation	Style of activity
Earthwatch	NPO	International	Environmental issue	All	Medium	Collaborative
Rio Tinto	BUS	International	Environmental issue	All	High negative	
The Prince's Trust	NPO	National	Social issue	All	High	Collaborative
Royal Bank of Scotland	BUS	National	Social issue	All	Neutral	

activities (international/national); (2) the purpose of the partnership (focusing on an environmental or social issue); (3) type of resources exchanged across the partner organisations (financial/non-financial); (4) the type of organisational reputation (a combination of three level scales of high-medium-low and positive-neutral-negative were employed based on the media comments and general perceptions); and (5) the style of activity among the two organisations which was constant (collaborative interaction) since the issue under examination was a collaborative relationship. Table 1 above summarises the different criteria employed.

The in-depth interviews were conducted with all the relevant people in each partnership case study and across both partner organizations. Semi-structured interviews were employed, in particular "problem centred interviews" which according to Flick (1998) are used in order to obtain subjective viewpoints about a social problem. After the transcription of the interviews the informants confirmed their interviews marking sections as 'off the record' (in addition to sections noted as 'off the record' during the interviews). Such 'off the record' requests were granted in full. For the purposes of the research including the identities of the organisations was an important part of the analysis due the constructs that were employed. Although different levels of employees were interviewed all the job titles of informants were replaced by the generic 'Interviewee' in order to mask the identity of informants. The final transcripts were imported into NVivo which was used to manage the data analysis process. A total of 837 nodes were initially developed and gradually collapsed into common themes grouped around the chronological stages of partnerships. Within the two case studies the aim was to arrive at theoretical rather than statistical generalisations (Ragin 1991) and to develop critical thinking (Alvarez et al. 1990) on partnerships. The analytic framework adopted was that of a contextualist approach following Pettigrew (1985) which highlights the importance of studying organisational change in three dimensions: context, content and process. All descriptions used for the conceptualisation of stages and processes are deeply grounded within the interviewees' comments using in most cases their own words. Initially narratives were created for each case study by combining the interviews, documentary evidence and secondary sources such as media commentaries.

Each case study narrative was sent to the case study organizations to ensure respondent validation (Bryman and Bell, 2007). The reference to dates and other historical information such as annual reports and other internal documents reflects the reality at the time of the data collection and analysis.

The case study organisations that participated in the research were for the first case study: Rio Tinto, Earthwatch, Partizans (a total of 16 interviews); for the second case study: Royal Bank of Scotland, and he Prince's Trust (a total of 24 interviews); and for the comparative interviews: Aviva, WWF, Shell, BP, BHP Billiton, Anglo-American, Dow Chemicals, BAA, Coca-Cola, FTSE, Deutsche Bank, HSBC, BT, MTV, BDS, Green Street Berman, Independent Consultant, CSR Network, Loop, Stop the Esso Campaign, Friends of the Earth, A&B, Platform, IPRA, World Business Council for Sustainable Development, International Textile, Garment & Leather Worker's Federation, CBI and Greenpeace (a total of 29 organisations and a total of 35 interviews).

The validity (Gibbs 2002) of the research has been ensured by a three stage feedback from the informants: (1) during the interviews; (2) after transcription; and (3) after the analysis of each case study. Furthermore, full copies of the final analysis were sent to the main informants of all the case study organizations. Also, all four case study organizations were offered the opportunity of a presentation of the main research findings at their premises. One such presentation was delivered with representatives of two case study organizations. The internal validity has been achieved through the consistent analysis of each case study and through the triangulation of the findings with the comparative interviews. 'External validity' refers to the extent to which a study's findings are capable of being generalised from the immediate case studies (Yin 1994:34). In case-oriented comparisons, "the goal of appreciating complexity is given precedence over the goal of achieving generality" (Ragin 1987:54). Nevertheless, in case study research generalisation is analytical rather than statistical, i.e. theories are tested through replication or contrasting results for predictable reasons (Yin 1994:34). Therefore, the strategy of data analysis followed the logic of "pattern-matching for rival explanations" (Yin 1994:108), i.e. the comparison of empirical results with (rival) patterns predicted in the literature, and "explanation building". The results were analysed and interpreted first within each case, and then compared across cases following the replication mode for multiple cases (Yin 1994:120). Finally, the findings of the case studies were corroborated with the findings of the 35 comparative interviews across 29 organizations. Concerning the reliability (Miles and Huberman 1994; Yin 1994:37) of the research the following measures were taken in order to attain maximum reliability: (a) multiple readings of transcriptions; (b) multiple readings of annual reports and other documents consulted; (c) use of interviews protocol and database; (d) detailed analysis of each interview separately; (e) analysis of summarised findings within each case study. The variety of groups and individuals interviewed allowed verification of the accuracy of information, and helped to put the respective views in perspective in order to achieve the study's aim to establish a chain of evidence showing the impact of partnerships based on the actors' perspective.

The study's time spans the four years of data collection and analysis i.e. 2002–2006. Hence, when reference is made to the time through the use of words such as 'current', 'at the time' and so forth, the above period is assumed as the current time. The choice of keeping the original time of the data collection and analysis is because it allows the reader to place the historical events within context according with the material at hand at that time. As organisations evolve through time the reality presented here refers only to the past, hence it does not take under consideration any recent changes, but rather reflects the original primary data collection.

Chapter 1
The Partnership Society

1.1 Introduction

The first chapter aims to introduce the main aspects of cross-sector partnerships, to present the working definitions within the book, the background trends that surround the relationship between nonprofit organisations and businesses and to situate the phenomenon under study within its historical context of interactions across the sectors.

The chapter sets the boundaries of the study by discussing the phenomenon of Nonprofit (NPO) Partnerships with Business (BUS) and its implications for society. It discusses the roles of societal sectors briefly in order to situate the study between the profit and the nonprofit sectors. It also offers an organisational/historical discussion of the relationships between the profit and nonprofit sectors in order to locate partnerships across the different types of associational relationships. Further on, it suggests the macro and meso forces that enabled the emergence of the partnership phenomenon on the global level and within the national context of the UK.

1.2 Societal Sectors, Roles and Definitions

"Institutions are the fundamental arrangements through which societies seek to deal with social and economic problems" (Weisbrod 1998:69). The triadic divide of society that started crystallising in the 1990s (Howell and Pearce 2001) is the predominant model employed in the literature (Seitanidi 2005). The public, profit and nonprofit model is an extension of the former divide between the state and the market which through the years "reached an impasse" (Weisbrod 1998:1). Each of the three sectors define their institutional plane in developed societies as presented in Fig. 1.1.

M.M. Seitanidi, *The Politics of Partnerships: A Critical Examination of Nonprofit-Business Partnerships*, DOI 10.1007/978-90-481-8547-4_1, © Springer Science + Business Media B.V. 2010

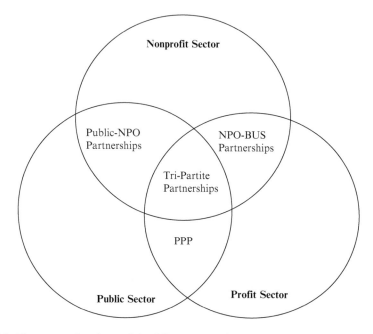

Fig. 1.1 The sectors of society and the different types of partnerships

The literature assumes distinctiveness between each sector based on their different roles. However, the distinctiveness becomes a little less firm if we look at the division of labour, which reveals the interconnections and interdependences among the sectors (Powell and Clemens 1998). The traditional role of the state is the welfare provision for its citizens. One of the roles of the public sector is to legislate and regulate in order to provide guidance and governance to and for societal actors. According to Black (2002:20) regulation is "a sustained and focused attempt to alter the behaviour of others according to defined standards or purposes with the intention of producing broadly identified outcomes and involving mechanisms of standard-setting, information-gathering and behaviour modification". Similarly one of the roles of the nonprofit sector, according to Tully (2004:1), is to develop rules "to control or influence corporate behaviour", which based on the definition provided by Black (2002) consists of regulation. Hence, the previously typical role of the state as a regulator is shared partially with the nonprofit sector.

Among the other more traditional roles of the nonprofit sector are to represent the interests of diverse groups and guarantee civil liberties, highlighting the political role of the sector (Meyer et al. 1997; Lewis 1999a) and to act as economic actors with "multiple bottom lines" (Anheir 2000) through the provision of funding for causes that serve the interests of the public at large (Tully 2004) but also as significant providers of employment (Salamon et al. 1999), highlighting the economic role of the sector (Anheir 2000).

The profit sector's fundamental role, on the other hand, is "the profitable production, distribution and sale of the goods and services. Other institutions will produce, distribute, and even sell goods. But, the notion that they do this in order to make a profit distinguishes commerce from other agencies" (Cannon 1992:36). Business is the institution that is best at "wealth and profit creation" (ibid). However, the public and the nonprofit sectors realised that both the skills and resources the profit sector possesses can be deployed more widely "in effect, the private sector might be better at performing some of the roles currently undertaken by the state or other agencies" (Cannon 1992:37). The main criticism directed to the profit sector with regard to the deployment of their resources through associational activities with the public and nonprofit sectors in order to serve the public good, is that they do not have the democratic mandate, the legitimacy or the historic role to act within this sphere and make socially justifiable decisions. Furthermore lobbying consists of the interface between the private sector and regulation. Paraphrasing Black's definition above on regulation it can be suggested that lobbying is also a sustained and focused attempt to alter the behaviour of regulators according to purposes that are served by the private sector. Hence one of the ways the private sector determines regulation is through lobbying. However, the public interest is not served by the interests of the private sector, which consequently remains to be safeguarded by the public and nonprofit sectors, or in other words the non-commercial sector (Seitanidi 2005).

The tri-sectoral divide of society is currently the predominant societal model employed and hence it will be assumed for the context of this book. The three sectors public, profit and nonprofit co-exist. In Fig. 1.1 the overlaps among the distinctive spheres of the three sectors can be observed. Although defining the overlapping areas is beyond the aims of the book, for the purpose of clarity providing examples can shed some light on these areas. Social entrepreneurship, for example, could be placed at the overlapping area between the nonprofit sector and the profit sector. Another example is the universities, which could be positioned at the overlapping area between the public and the nonprofit sectors. Figure 1.1 portrays the different types of partnerships and where they are positioned with regard to the three societal sectors: firstly, PPP (Public–Private Partnerships) are between public and private organisations; secondly, Public-NPO Partnerships are between public and nonprofit organisations; thirdly, nonprofit and business partnerships (NPO-BUS Partnerships) are the type of partnerships this book is examining. Fourthly, tripartite partnerships involve all three sectors.

Over recent decades, the relationship between the market and society has been going through transformation. The process of commodification and liberalisation of markets in developed economies has been central to the advancement of capitalism and has resulted in a 'market society'. Corporations have assumed a role of greater importance, as powerful actors controlling a multitude of resources and eager to attain access not only to financial but equally to non-financial resources 'intangible aspects' (Millar et al. 2004), which they gradually revaluated as equally if not more important than the financial ones (Seitanidi 2007a). On the other hand, the gradual empowerment of civil society organisations has played an important role in transforming nonprofit entities into equally powerful forces of change within society

(Doh and Teegen 2002; Bendell 2000c). The role of the state is also being re-defined (Mol 2007) while its limits and desirability are being questioned.

The roles and responsibilities of each sector are not clear as new flexible forms emerge for the delivery of goods and services, such as those of partnerships:

> there have also been crucial changes in the status of civil society as a partner, as most developed societies move towards a situation in which all three sectors share the role of providing public goods, if in different constellations in different circumstances. (Deakin 2002:136)

Partnership is the dynamic constellation of entities across different sectors that can attempt to provide society with 'public goods', such as clean water, clean air, health care, and education.[1] Although in the past the provision of safe, just and trustworthy public goods was considered the ethical monopoly of the state, today corporations and NPOs are playing an equally important role in shaping the agendas of government (Crane and Matten 2004). Elected governments represented the will of the majority of citizens who exercised the system of representative democracy. However today, given the decrease in political participation the model of participative democracy does not equate with the majority of the citizens. The politics of the 'strong minorities', i.e. those who control the major resources, are in operation. Even if it was common knowledge that companies, or indeed other actors, were determining the agendas of governments (Garney 1998), in the past through either legal or non-legal forms of direct or indirect governance, the ultimate responsibility of outcomes still fell to the state. Partnerships can be seen as a new micro-associational domain in which the roles and responsibilities are negotiated through the participating organisations. Each participating entity represents not only its own rights and responsibilities in the specific situation, but also those of its sector, through the implications following the partnership relationship. Since partnerships are attempting to solve complex social problems that the state or individual sectors were unable to solve or chose not to address, the importance and implications of the politics of this process are crucial for the new agendas that are being formed. In fact, partnerships have replaced direct intervention and legislation as the two main types of government reaction to problems (Margaret Hodge, MP, Under-Secretary of State for the Department of Education and Employment, 2000, citied in Jupp 2000:11).

In effect, politics in this book refers to the process of decision making within the micro-associational domain of a partnership relationship between a profit and a nonprofit organisation. Political science focuses on "power and the distribution of power in different political systems. Political scientists enquire as to the sources of power, how it is exercised and by whom, how processes of constraint and control operate, who gains and who loses in power struggles" (Scott and Marshall 2005). The narrow view of politics refers to the political institutions such as the state, political parties and interests groups that play a role in the process

---

[1]For cases of these types of partnerships see: Warner and Sullivan 2004.

of policy making (ibid). The broad view of politics on the other hand examines the operation of power beyond the political institutions outside the political system and involves for example the power of corporations within capitalist economies (ibid). Calhoun (2002) suggests that

> "ancient Greek thought and its heirs also stress a view of politics as constitutive of legitimate social life, insofar as it provides ways for individuals to express their opinions, influence each other, and build institutions. ... Defining what counts as political has always been inseparable from the question of how societies govern themselves. Human life is fulfilled through active participation in the definition and construction of the social world – quintessentially political activities-and that this process involves recognising the perspectives of others and experiencing, at first hand, the interdependence and cooperative dimensions of society" (ibid).

Partnership is today a widespread social phenomenon that constitutes a process of joint decision-making in a number of spheres that were previously considered part of the public policy realm (Waddock 1988). Consequently partnerships are seen as a new system that societies use to govern themselves (Glasbergen 2007) through the participation of different economic sectors, the perspectives they represent reflecting the interdependence of the sectors and organisational actors.

As a result, partnerships today address the previous criticism that was directed to the profit sector with regard to their mandate to serve the public good. Since decisions regarding the corporate involvement in society are taken 'in partnership' with the nonprofit sector (and with government in the case of tripartite partnerships) that has the mandate (even not democratically at all times) to serve the public good, then it appears that the criticism cannot any more be raised with regard to the legitimacy of the profit sector.

The rhetoric around partnerships presents this collaborative associational form as one of the answers to the complex problems society is faced with today (Birch 2003; Warhurst 2001); hence all three sectors are asked to address these complex problems by employing all available resources. A partnership, however, as a non-regulated form of association, allows for the diffusion of responsibilities to all social actors and sectors involved. If all three sectors of society are involved in a tripartite sector partnership, for example, then in the event that an outcome is not satisfactory to the people affected by the solution, it becomes problematic to hold any one of the actors accountable for the process or outcome. Furthermore, the delineation of roles and responsibilities of each sector involved in a partnership as well as the outcomes of the relationships are a relatively new area of examination within the NPO-BUS partnership literature (Seitanidi 2008; Biermann et al. 2007; Selsky and Parker 2005).

Moreover, the literature has paid less attention to the tensions between NPOs within the broader civil society and the market (Howell and Pearce 2001). The reason for this is that these tensions exist between the two different spheres of reality (civil society and the market) that represent also distinctive research areas and disciplines (social policy, nonprofit management, sociology and the broad area of business management, including business ethics, corporate social responsibility). However the tensions within NPO-BUS partnerships, for example, exist in this mid-section space. The management literature predominately employs a functionalist perspective, hence these tensions are either being ignored or seen as dysfunctional hence utilise conceptualisations of strategy in order to appropriate them and forge consensus. On the other hand, within social policy and nonprofit management the studies that focus on these tensions predominately focus on developing countries (Howell and Pearce 2001; Lister 2001). Another reason is "the tendency to assume that civil society within nation states is homogeneous in moral purpose and values" (Howell and Pearce 2001:118) and hence extent the assumption to NPOs. This in effect underplays the need to examine the tensions that exist in retaining the conflictual elements of the nonprofit sector in its relations with the state and the market. This study will attempt to close this gap in the literature by focusing on the tensions between NPOs and BUSs in a developed and mature economy.

1.2.1   Definitions of Terms

It is important to discuss the terms that employed within the literature in order to justify their selection but more importantly to develop a definitional basis that will assist in the empirical chapters of the book.

1.2.1.1   Nonprofit Organisations (NPOs)

The generic term NPO is used in order to refer to all the different types of nonprofit organisations. The different phrases employed to describe the nonprofit sector, such as 'voluntary sector', 'charitable sector', 'independent sector' emphasise the different attributes that characterise nonprofit organisations (NPO). The distinction between the different terminology utilised is also a reflection of the "parallel research universes" (Lewis 1999a:1) in the literature as well as the different types of organisations under study and the issues researched (Mowjee 2001). More specifically the term 'voluntary organisations' is used in the literature mainly to describe nonprofit making organisations with a particular interest in welfare activities in industrialised countries. On the other hand the term 'non-governmental organisations' (NGOs) is used primarily to denote nonprofit making organisations involved in development and relief activities in aid-recipient countries (Mowjee 2001; Lewis 1999a; Billis and MacKeith 1993). Finally, the voluntary organisations literature is typically focused on explanations for the existence of the sector, organisational and management issues, and the changing policy context in which the organisations operate. The main concentration of the NGO literature is on the role of NGOs in development, relief and social change and on development practise and NGO relations with states and donors (Lewis 1999a).

As Salamon and Anheier (1997:33–34) argue, the organisations that represent the sector have sufficient similarities to apply a 'structural operational' definition. More specifically, they define the nonprofit sector as:

A collection of entities that are organised (i.e. institutionalised to some extent; thus ad hoc, informal or temporary gatherings of people are not considered part of the nonprofit sector), private (i.e. institutionally separate from the government), nonprofit distributing (i.e. not returning any profits generated to their owners or directors), self governing (i.e. equipped to control their own activities) and voluntary (i.e. involving some meaningful degree of voluntary participation, either in the actual conduct of the agency's activities or in the management of its affairs).

The above definition employs only structural operational characteristics in describing the nonprofit entities and as pointed out by Anheir (2001:4) "definitions are neither true nor false. They are judged by their usefulness in describing a part of reality of interest to us." However, as the aim of the book is to study the phenomenon of NPO-BUS partnerships, it takes the view that partnerships evolve within a 'social space' or 'social eco-system', where organisational actors negotiate costs and benefits not only for themselves but at the same time for society at large, since they are interdependent and interconnected entities. Consequently, the partners involved are: the business organisation, the nonprofit entity and society. Society, as a silent partner, mainly participates in the partnership through the nonprofit entity. As the Preamble to the Statute of Charitable Uses of 1601 in England describes, the most common type of function of the nonprofit sector is the promotion of what is termed 'public interest'. The element of public interest in NPO is best represented in the term 'civil society' as argued by Edwards and Gaventa (2001:2):

civil society is a contentious term with no common or consensus definition. It is the arena in which people come together to advance the interests they hold in common, not for profit or political power, but because they care enough about something to take collective action.

The Centre for Civil Society (CCS 2006) at the London School of Economics defines civil society as:

Civil society refers to the arena of uncoerced collective action around shared interests, purposes and values. In theory, its institutional forms are distinct from those of the state, family and market, though in practise, the boundaries between state, civil society, family and market are often complex, blurred and negotiated. Civil society commonly embraces a diversity of spaces, actors and institutional forms, varying in their degree of formality, autonomy and power. Civil societies are often populated by organisations such as registered charities, development non-governmental organisations, community groups, women's organisations, faith-based organisations, professional associations, trades unions, self-help groups, social movements, business associations, coalitions and advocacy group.

The reason that the term 'civil society organisation' (CSO) is usually employed in the literature is that it is deemed closer to the issues of legitimacy, representation and accountability. The importance of legitimacy[2] is captured by the question 'Who

---

[2]As contended by Edwards and Gaventa (2001:7) "Legitimacy is generally understood as the right to be and do something in society – a sense that an organisation is lawful, admissible, and justified in its chosen course of action. However, there are generally two ways to validate an organisation: through representation (which usually confers the right to participate in decision making) and through effectiveness (which only confers the right to be heard)."

speaks for whom?' and refers to the process of representation and accountability in the establishment of the nonprofit organisations; e.g. the extent to which NPOs are closely identified with the citizens and their concerns/interests or the issues they are supposed to represent. Furthermore, the term 'civil society organisations' captures the attribute of active participation of the 'civil' society, to some extent ignoring of the 'uncivil' society. Successively, when referring to civil society organisations the definition below is used:

> Civil society organisations are all those bodies that act in this arena (civil society as defined above), comprising a huge variety of networks and associations, political parties, community groups, and NGOs but excluding firms that are organised to make a profit for their shareholders and that generate no public benefits." (Edwards and Gaventa 2001:2)

However, it seems that in reality the terms 'civil society', 'nonprofit organisations' or 'NGOs' are used interchangeably, depending on the frame of the discussion and primarily to evoke the connotations associated with each of the terms. For example, NGO is used to refer to the radical nonprofit organisations (such as Greenpeace and Friends of the Earth). It appears that the term is primarily used symbolically to manifest certain aspects of NGOs, such as representation (membership-based organisations), their tactics and organisation identity characteristics.

This study involves a wide range of nonprofit organisations (NPOs), NGOs and CSOs. In cases where a different term is used within the study, rather than the generic term NPO, the reason is to associate commonly held connotations to the organisations discussed. The employment of each term discussed above within the book entails the definitions provided in this section.

1.2.1.2 Business (BUS)

Business refers to a collection of individuals and structures grouped under the legal form of corporation[3] in order to increase and maintain profit within particular spheres of interest. Since the eighteenth century the private pursuit of profit has increased in importance (Solomon 2004). Parallel to the prominence of attainment of profit through the operations of a business sprang the increase in the misuse of corporate power due the original "divorce of ownership and control" (Solomon 2004:3). As a result the responsibilities that used to reside only with the owner(s) of a corporation who also had the control of the business where separated and dispersed among shareholders and management. The effects of the above were beneficial for business profitability but in many cases detrimental for wider parts of society due to their interface with the corporate greed; examples include: over-priced stock price, excessive executive remuneration, selling knowingly unhealthy or dangerous products; destroying the environment; unfair treatment of employees and so forth. Hence when any of the above instances of human greed would surface on the public sphere corporations were confronted with reputational crisis which

---

[3]For a legal and historical account of the development of corporations see Farrar and Hannigan (1998).

impacted with the public's perception of their role. Such crises gradually gathered momentum resulted in cynicism towards business which according to Trevino and Nelson (2007:3) "has become an epidemic throughout society". While the influence of business has been increasing (Solomon 2004) similarly the demands for an institutionalised level of responsibilities has found voice in the conceptualisation of corporate social responsibility (CSR) (Crane and Matten 2004). CSR is a call for business to operate in a responsible way which appeared as an antiphon to the divorce of ownership and control aiming to reform the practises of business by introducing an organised and institutionalised response from within corporations.

Corporate Social Responsibility (CSR) is today a worldwide phenomenon which has been encouraged by business (McWilliams and Siegel 2002; Zadek 2001), by nonprofit organisations (Bendell and Lake 2000) and by government (Moon 2004; Seitanidi 2008). The European Commission's Green Paper (Commission of the European Communities 2001) defines CSR as: "a concept whereby companies integrate social and environmental concerns in their business operations and in their interaction with their stakeholders on a voluntary basis". Moon (2002) argued that the network mode of operation of business with government and NPOs represented a new system of re-orientation of governance roles among the sectors whereby the increased interdependencies were guided by the pursuit of shared interests and values. He suggested that in the UK "CSR was part of a wider re-orientation of governance roles" (Moon 2004:1). The case of the UK has a particular interest as CSR has been "institutionalised" which according to Moon (2004:22) "has been in large part a function of government which, in turn, has sought to respond to governance deficits". One example of the level of institutionalisation is that the UK was the first country to have a dedicated Minister for CSR. The institutionalisation of CSR has stimulated further discussions and actions concerning the responsibilities of each sector but also in attempting to reach solutions for the wider social and environmental issues by implementing new flexible forms of governance.

Corporate social responsibility has been characterised as "an eclectic field with loose boundaries, multiple memberships, and differing training/perspectives; broadly rather than focused, ..." (Carroll 1994:14). Carriga and Mele (2004) suggested that all relevant CSR theories can be included in four groups: instrumental, political, integrative and value theories, each one pointing to a particular way to perceive the role of business in society in turn: "(1) meeting objectives that produce long-term profits, (2) using business power in a responsible way, (3) integrating social demands and (4) contributing to a good society by doing what is ethically correct." (ibid 66). Within the above four groups there are a number of approaches such as sustainable development (Gladwin and Kennelly 1995), normative stakeholder theory (Freeman 1984; Donaldson and Preston 1995), corporate citizenship (Matten and Crane 2005; Wood and Lodgson 2002), stakeholder management (Rowley 1997; Mitchell et al. 1997), corporate social performance (Wood 1991; Wartick and Cochran 1985) to mention only a few.

In practise the arguments that support the implementation of corporate social responsibility (CSR) framework appear victorious; unlike the theoretical debates that still continue to examine the arguments 'for' and 'against' CSR as a concept

(Carriga and Mele 2004), or the long running debate about whether CSR pays off (Schuler and Cording 2006; Mir et al. 2006; Vogel 2005). Quite simply, organisations faced with CSR problems and challenges need effective ways of implementing CSR programmes and initiatives regardless of whether CSR is 'right' or whether CSR leads to better financial performance. The triumph of voluntary CSR implementation in practise is testified by the wide array of initiatives that range from indicators for responsible practises in the stock exchange markets[4] to social and environmental reporting[5] and a multitude of corporate community involvement programmes. However, whilst there is an emerging consensus that CSR can and should be implemented in organisations, "CSR is currently characterised by many unsystematic practises, i.e. constellations of arrangements that are fit for purpose within specific contexts but which lack transferability and sustainability" (Seitanidi and Crane 2008:251). Cross-sector partnerships is one of the most exciting, promising yet challenging ways that organisations employ to implement CSR in recent years (ibid).

1.2.1.3   NPO-BUS Partnerships

In the context of this book the term NPO-BUS (nonprofit and business) Partnerships will be used to refer to the type of partnerships under study. There are different ways to describe these types of partnerships such as 'cross-sector partnerships' or 'social partnerships'. Waddock (1988:18) defines social partnerships as:

> A commitment by a corporation or a group of corporations to work with an organisation from a different economic sector (public or nonprofit). It involves a commitment of resources – time and effort – by individuals from all partner organisations. These individuals work co-operatively to solve problems that affect them all. The problem can be defined at least in part as a social issue; its solution will benefit all partners. Social partnership addresses issues that extend beyond organisational boundaries and traditional goals and lie within the traditional realm of public policy – that is, in the social arena. It requires active rather than passive involvement from all parties. Participants must make a resource commitment that is more than merely monetary.

It is argued by Googins and Rochlin (2000:131) that although Waddock's definition captures successfully the wide spectrum of what constitutes a partnership between

---

[4]In the London Stock Exchange FTSE4GOOD is a special index that includes companies that meet the inclusion criteria that offer testimony of responsible practices. For more information: http://www.ftse.com/Indices/FTSE4Good_Index_Series/Criteria_Documents/index.jsp
Also in the US the Dow Jones Sustainability Index (DJSI), launched in 1999, a global index that aims to track the financial performance of sustainability-driven companies worldwide and currently manages over 4 billion euros. For more information:
http://www.sustainability-indexes.com/htmle/other/faq.html

[5]In 2005 the GRI reported 750 companies that used the Sustainability Reporting guidelines (GRI 2006). The Global compact reported that world wide (last update of information on line 29 March 2006) 2,500 businesses are included in its network (Global Compact 2006).

the "socially-driven and market-driven entities", it does not offer answers as to what motivates the partners. Wilson and Charlton (1993:10) in their definition move towards the motivational issues involved in the partnership:

> It has been suggested that a partnership should seek to achieve an objective that no single organisation could achieve alone – an idea described by Huxham (1993) as 'collaborative advantage'. This is a common concept in business where strategic alliances and joint ventures are only entered into when there is added value to be derived from organisations working collectively… (T)he risks and benefits of the venture need to be shared, so when success is achieved all partners are better off. This implies that there needs to be a degree of mutuality of benefits across partner organisations.

Although the aforementioned definition provides an indication of where the motivation for the partners might lie, it lacks focus in explaining the different types of motivations that organisations might have. Each partner represents a different perspective (Holzer 2001; Austin 2000; Kanter 1999). The extent to which the motivations are shared among the partners has not been examined in depth, which is one of the aims of this study. Furthermore although the above definition suggests the need for a collaborative advantage that the partnership relationship should deliver to both partners, it lacks the social dimension that Waddock's definition suggested previously. This research examines the extent to which the positive outcomes delivered by a partnership relationship extend to the social domain and if indeed they are the result of combined efforts between the partners. More importantly it aims to qualify whether the relationships under study are or are not social partnerships based on their delivery of positive outcomes to society, as suggested by Waddock (1988). It is within this context that the study aims to connect positive partnership outcomes for the benefit of society with sustainability. Sustainable development is the most prominent discourse of ecological concern (Dryzek 2005). Despite the difficulties in defining sustainable development for the last 20 years (Dobson 1998; Meadows et al. 1992; Barbier 1987) Brutland's definition prevailed as the most widely quoted: "Humanity has the ability to make development sustainable-to ensure that it meets the needs of the present without compromising the ability of future generations to meet their own needs" (World Commission on Environment and Development 1987:8). The fundamental assumption of sustainability is the need for change (Glasbergen 2007; Dryzek 2005) on the local, national and global levels as the above report explicitly proclaims: "In essence, sustainable development is a process of change in which the exploitation of resources, the direction of investments, the orientation of technological development, and institutional change are all in harmony and enhance both current and future potential to meet human needs and aspirations" (World Commission on Environment and Development 1987:46). As Dryzek (2005:146) points out "Brutland's definition did not satisfy everyone, and other definitions of sustainable development proliferated "opinions differ as to what human needs count, what is to be sustained, for how long, for whom, and in what terms". The answers to the above vary depending on the political interpretation of different interests that are being represented or not as the case might be; hence Dryzek

(2005:147) refers to sustainable development as a 'discourse' rather a concept. Sustainable development was introduced as a continuum by the Brutland report as it tried to combine previously conflicting notions such as international development, environmental issues, social justice and economic growth, however failing to provide an implementation plan (ibid). In effect corporations were not any longer the problem but were accepted as part of the solution. This new role of business "was solidified in partnerships involving business, government and NGOs several hundreds of which were established at the WSSD" (ibid 150) (World Summit for Sustainable Development) or as some refer to them "the privatization of sustainable development" (Von Frantzious 2004:469). Since the WSSD in 2002, Type II Partnerships,[6] particularly within the environmental sector, came to the forefront as the optimum way of 'making a difference'. One of the requirements for sustainable development is the shift in the power (Dryzek 2005) between different actors on the political, economic and societal levels. Not all partnerships allow for the shift in power to take place and to arrive at sustainable solutions. Although the motivation behind sustainability discourse can be present, the outcomes are not always within reach. Sustainable solutions are seen as the result of change processes in institutional, organisational and individual agents and the delivery of positive outcomes to society as a result of partnership relationships that extend beyond the strict criteria of Type II Partnerships, i.e. they do not focus on the sustainable development agenda originally promoted at the WSSD. The aim is to associate sustainability with the outcomes within the broad context of NPO-BUS Partnerships in order to move away from the labyrinth of good intentions to the avenue of outcomes that will deliver sustainable solutions. In effect the book not only critically examines partnerships but presents a study of the changes that occur in organisations as a result of partnership relationships discussing the types of positive outcomes that are generated.

Figure 1.2 presents the three types of dual social partnerships that exist: the Public-Private Partnerships, the Public-Nonprofit Partnerships and the Private-Nonprofit Partnerships, that the book is concentrating upon. Also there is a tripartite type of partnership (TPP) that involves all three sectors (Warhurst 2001). Finally, Business-Business Partnerships (Kanter 1994) are a further type of partnership, but one which is not classified as a social partnership as the aim is not to address a social issue.

Before we examine the literature on NPO-BUS partnerships it is important to position partnerships between the two sectors within the historical/organisational context of collaborative relationships. This is discussed in the subsequent section as depicted in Austin's (2000) collaboration continuum between the nonprofit organisations and businesses.

---

[6]Type I partnerships refer to the negotiated agreements between governments that aim to implement the aims of Agenda 21. Type II partnerships are non-negotiated informal agreements between governments, intergovernmental agencies and civil society actors that aim to complement the negotiated 'Type I' commitments.

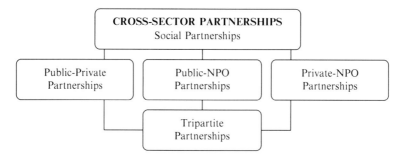

Fig. 1.2 Cross-sector social partnerships (Redrawn from Seitanidi 2007)

1.3 The Collaboration Continuum

In the late 1990s the idea of cross-sector partnerships was relatively new in Europe. When the term 'partnership' was employed it was primarily used to refer to partnerships between the government and businesses usually termed as 'PPP' (Public–Private Partnerships). On the other hand the term 'strategic alliances' was used to describe business-to-business partnerships (Hemphill and Vonortas 2003; Koza and Lewin 1998; Kanter 1994; DiMaggio and Powell 1983; Astley and Fombrun 1983). Socio-Sponsorship[7] still prevailed as the most practised associational form within the corporate involvement sphere. There was no clear link between sponsorship and partnership apart from the fact that both were associational forms between the two sectors.

James Austin, of the Harvard Business School, was the first academic (Austin 2000) to offer a conceptual framework for the emergence of the NPO-BUS partnership phenomenon. More importantly he positioned previous forms of associational activity between the profit and the nonprofit sectors in a continuum. In other words, he placed philanthropy and sponsorship in the same continuum as partnerships. This was an important conceptual contribution, as it allowed for a systematic and cohesive examination of previously disparate associational forms. The "Collab oration Continuum" is a dynamic conceptual framework that contains two parameters of the associational activity: the degree, referring to the intensity of the relationship, and the form of interaction, referring to the structural arrangement between nonprofits and corporations (ibid 21), which he based on the recognition that cross-sector relationships come in many forms and evolve over time. In fact, he termed the three stages that a relationship between the sectors may pass through as: philanthropic, transactional and integrative.

---

[7]As defined by Coutoupis (1996:24): "Sponsorship (social) is the financing and support of nonprofit organisations and/or activities of social context, from private businesses, with the exclusive compensation the credit of sponsors with social beneficence, comprising the transfer of resources from the private to the public-social sector".

Austin's theory (Austin 2000:35–37) suggested that we have moved from philanthropic relationships, where philanthropy monopolised the relationships between the two sectors (philanthropic stage) and where financial resources were transferred from the profit to the nonprofit sector, to transactional relationships, including cause-related marketing (CRM), and sponsorships (transactional stage). In the transactional stage the magnitude of resources changed the emphasis to the organisational mission and the level of managerial involvement increased. Today, according to Austin, a higher associational form has emerged, termed 'integrative relationships' where partnerships are positioned. In partnerships, according to Austin (2000:35–37), there is a joint benefit and value creation, a higher level of management involvement, where the importance of the relationship to the mission of the organisation changes from being peripheral (in the philanthropic stage) to being strategic (in the integrative stage). Instead of being narrow, the scope of activities becomes broad, the managerial complexity increases and the resources exchanged are big and multiple. One of the important aspects that Austin points out is the collaboration mind-set which in the integrative stage is expressed with the 'we' mentality instead of 'us versus them'. The Collaboration Continuum is a dynamic process, according to Austin, which can evolve incrementally from one stage into another but equally can regress to previous stages. As Austin points out "the three stages are not single discrete locations; there are many points in between the stages" (Austin 2000:35). In particular transactional relationships might not be preceded by philanthropic stage relationships and might be the initial stage for a relationship between the two sectors.

Using Austin's descriptions, Table 1.1 was devised with a view to offering comparisons among the three different relationship stages, which are not explicit in Austin's tables and figures in the original publication (Austin 2000:35–36).

Austin predominantly pays attention to the strategic aspects of the collaboration and only makes reference to the social outcomes. He refers to the programmes that were devised during the partnership relationships, based on the case studies he examined in his book within the American context. He does not appear to qualify partnership relationships on the basis of their social outcomes, as Waddock (1988) suggests in her definition. His work, among others (Warner and Sullivan 2004; Birch 2003; Elkington and Fennell 2000; Andrioff 2000), marks a shift in the literature reflecting the current practises in the field of stirring these relationships towards a strategic approach that aims predominately to deliver positive outcomes for the two partners. Although the social character of partnerships has been discussed in the literature, it has not been examined in detail within partnership case studies. This strategic shift in partnerships increases the value of partnerships for the partners but underplays the role of these relationships for society. This issue will be further discussed as part of the literature review in the next chapter.

The above section discussed NPO-BUS Partnerships within the historical context of the interactions across the two sectors. The following part of this chapter will examine the factors that contributed to the development of the partnership relationships within both the global and the national UK context.

Table 1.1 The characteristics of the collaboration continuum stages (Based on Austin's descriptions (2000:35–36)

	Philanthropic stage	Transactional stage	Integrative stage
Magnitude of resources involved	Not economically critical	More important resource exchanged (more than monetary)	Increase of resources exchanged
Level of importance to organisational mission	Not particularly important	Overlaps in mission and potential for value creation	Collective action and organisational integration
Level of benefits	Modest (for NPO increase in funding and for the BUS enhancement of its reputation)	Increase in strategic importance, connected more directly with business operations, core competence exchanged	Joint benefit creation, Projects identified and developed at all levels in the organisation, with leadership support
Level of managerial involvement	Low (no top level involvement)	Shared vision at the top of the organisations	Company's executive named to nonprofit partner's board of directors
Mode of engagement	Minimised interaction and communication	More active	Intensity in personnel interaction
Level of value flow	One-way	Significantly two-way	Individual value creation escalates to joint value creation

1.4 NPO-BUS Partnership Enablers: Macro and Meso Forces

This section of the chapter discusses the different contributing factors for the emergence of social partnerships globally but also within the national context of the UK more specifically. The macro forces refer to the global phenomena that enabled the two sectors to develop closer relationships. The meso forces refer to the contributing factors that appeared within the institutional environment of the UK.

On the global level and within the developed economies globalisation, the lack of international legislation, the decline of the old role of the state, the empowerment of business, the communications revolution, the empowerment of civil society organisations and the changes in consumerism are considered to be the macro forces that have influenced the phenomenon under study, and are discussed below (Fig. 1.3).

Globalisation is one of the forces that impact on the relationship between the businesses and civil society organisations. According to Giddens: "The simplest

---

Macro Forces

Globalisation
Decline of the Old Role of the State
Empowerment of Business
Communications' Revolution
Empowerment of Civil Society
Organisations
Changes in Consumerism

Meso Forces

National Devolution of Power
Fragmentation of Delivery
Crisis of Legitimacy
Deficit of Governance
Decline in the Levels of Trust
Increased Levels of Sophistication in Funding
Institutionalisation of CSR

Fig. 1.3 The macro and meso forces

---

definition of globalisation is interdependence, living in a more global world means living in a more interdependent world where events happening at one side of the world in a direct way impact on what is happening at any particular place in the world that one happens to be" (Giddens 2001). The "second global age" (Giddens 2001:4) in which we are currently living can be characterised as financial "2 trillion dollars turned over everyday on world currency markets, it's an extraordinary figure and it was only a fraction of that figure, 30 or so years ago" (ibid 5), also political, cultural and primarily bound up by communications. In fact, Giddens remarks that "The most important transformative force in our lives over the last 30 or so years is not economic markets, it's not economic interdependence, it is the impact of communications, especially, it is normally called the *communications revolution*" (ibid:5). The development of telecommunications, specifically of the internet, allowed civil society organisations to communicate their messages more widely, faster and with minimal costs involved. This had a profound effect in the NGO movement, since the internet was used, according to Bray (2000:50–53) as both "a source of information" and "a campaign tool". Civil society organisations were recognised as symbols of equal power with business, yet of a different type of power (Bendell 2000b). They started developing a higher level of sophistication in their strategies and tactics, gaining a newly-found respect for their brand identities from the business community resulting in the *empowerment of civil society organisations*.

Within the globalisation phenomenon as observed above we have the rise of parallel phenomena such as the *decline of the old role of the state* (Bendell 2000c), combined with deregulation and liberalisation (Newell 2000) of national economies, as Giddens also points out above. It is important to note though, as explained by Tarrow (2001):2), that "states remain dominant in most areas of policy - for example, in maintaining domestic security". The criticism that is associated with the role of the state is primarily directed to the continually diminishing ability of governments to control capital flows (Krasner 1995; Risse 2001; Spruyt 1994). Thus the *empowerment of corporations* (Crane and Matten 2004) as important financial actors who can also exercise political power makes more visible the insufficient flexibility of the state (Newell 2000) to offer solutions to problems.

Giddens (2001:6) remarks regarding the above: "Daniel Bell[8] had perhaps the most celebrated quote about this than anyone has made, when he said 'the nation state has become too small to solve the big problems and too big too solve the small problems'". Hence, we are witnessing a decline in the ability of the state to provide for social goods as those were defined and expected only a few years ago.

Changes in consumerism is another widespread phenomenon that impacted and still impacts upon the relationship of the two sectors. Chapter 4 of Agenda 21 commences with the following statement: "The major cause of the continued deterioration of the global environment is the unsustainable pattern of consumption and production, particularly in the industrialised countries" (Blaza et al. 2002:1). Also Principle 8 of the Rio Declaration states: "To achieve *sustainable development* and a higher quality of life for all people, Member States should reduce and eliminate unsustainable patterns of production and consumption" (ibid). Moreover the Ethical Purchasing Index (EPI) in the UK shows that there is a rapid growth of ethically-sound goods and services. The value of consumer ethical purchases across the index sectors grew from £4.8 to £5.7 billion between 1999 and 2000, an increase of 18.2%, compared to 2.8% for the economy as a whole." (Doane 2001). In fact, the EPI did not include ethical investments and banking, which account for an additional £7.8 billion (having a growth rate of 20% per year). This testifies to a shift in the consumption patterns, which is important not predominantly in monetary terms as yet but as an indicator of changes in people's decision-making processes in their everyday life and in the future. This shift could be seen by all three sectors, but primarily the profit and nonprofit sectors, as an opportunity for promoting their own agendas. It seems that due to the nonprofit sector's missions (especially within the environmental NPOs) this trend is closer to their own agendas as they have played an active role through the years in changing the patterns of consumption in developed economies (Bendell 2000b). Consumer politics similarly influence the purchases of consumers in order to exert control in companies and force them to change their production patterns (Bendell 2000b). Today firms have to deal with global protesters who by using the internet can organise faster all around the world (Crane and Matten 2004; Klein 2000).

On the other hand, the *meso-forces* are the national devolution of power, the fragmentation of delivery, the crisis of legitimacy, the deficit of governance, the decline in the levels of trust, the increased levels of sophistication in funding and the institutionalisation of CSR in the UK (Fig. 1.3). According to Giddens (2001) the UK has been going through a process of *national devolution of power* which is likely to continue in the future. In his opinion the devolution process interacts with globalisation enforcing the outcome: "Globalisation, because it pulls some powers away from the nation, it also pushes down, it forces decentralisation of power away from national government". In the 1980s the UK witnessed the government's

---

[8]Daniel Bell is a renowned sociologist.

attempts to decentralise its agencies in order to improve "efficiency and responsiveness", which resulted in a "single-focus approach [and which] exacerbated the difficulty of co-ordinating multi-agency responses to complex problems" (Ling 2002:618) hence it was characterised as the era of *'fragmentation of delivery'* (ibid).

Partnerships arose within the "joined up working" (ibid) initiatives and as an effort to address the system's previous limitations. Partnerships have been widely used in the UK as an alternative form of governance (Sullivan and Skelcher 2002; Moon 2004; Rowe and Devanney 2003; Miller and Ahmad 2000), a non-regulated form, where all sectors of society are trying to address a multitude of pressing problems. In fact, it seems that the partnership approach in the UK[9] was a *government-led concept* (ibid), not just by the current Labour administration but also by the previous Conservative administration (Moon 2004). The government encouraged companies as well as civil society organisations to participate in partnerships, which can be seen as a way to address the *'crisis of legitimacy'* of governments in the UK (Henn et al. 2000) which further added to the *deficit of governance*. According to Bovaird et al., (2002:421) "the high desire of local authorities to engage in partnership working with the voluntary sector, in order to hive off some of their responsibilities, especially for high costs and high risk services such as care of people with disabilities or mental health problems." In fact, according to Jupp (2000:17), 'since Labour's election to government in 1997, the emphasis on cross sector collaboration has accelerated. This trend is likely to continue, at least until it meets the limits of public tolerance'. An example of this trend was the Private Finance Initiative (PFI), which "required all public sector agencies seeking capital investment to try and obtain it via partnership with the private sector" (Sussex 2003:59). Even though the PFI was introduced by the Conservative Government, the Labour Government carried it through with similar enthusiasm.

The crisis of legitimacy is not an exclusively British phenomenon (Bovaird et al. 2002) and it refers to a "widespread distrust of big business, political parties, trade unions, the Church and the press" (ibid 427). Although "there are no Eurobarometer time series on the evolution of trust in institutions" (ibid), the 2004 data of the National Report of Eurobarometer 61 (2004) for the UK paint a picture of *the levels of trust* certain institutions enjoy. A number of institutions in Britain have the lowest levels of trust (Figs. 1.4 and 1.5) in the EU15, namely: the written press (20%), the national parliament (25%), the national government (19%) and the political parties (10%). In the following institutions trust levels dropped: the justice system (the UK trust numbers have dropped within the last 6 months[10]) to 37% (−6%)[11]; the police 55% (−5%); the army 67% (−5%); the religious institutions 37% (0%); the trade unions 34% (−1%); big companies 20% (−3%); charitable and voluntary

---

[9] It is similarly argued by Hardis (2003) that partnership discourse was cultivated by the Danish state.

[10] The latest Eurobarometers research consulted for this research took place in July 2004 and the previous one in October 2003.

[11] The number in the bracket notes the percentage decrease.

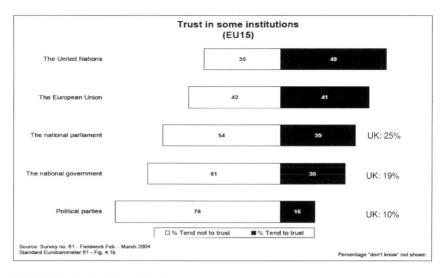

Fig. 1.4 Trust in institutions Part A

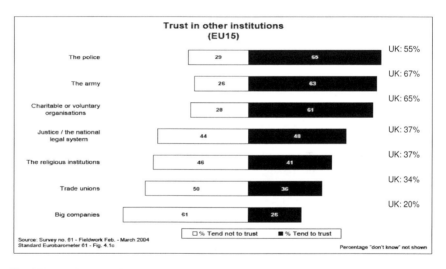

Fig. 1.5 Trust in institutions Part B

organisations 65% (−3%) (Fig. 1.5). As far as big companies are concerned, Sweden had the highest figure of mistrust, 68%, compared with the UK levels of 65%, with the European average at 27%. Overall it is apparent that the levels of trust decline in most institutions in the country.

The above picture testifies clearly why big businesses were interested in forming partnerships or indeed getting closer to the voluntary sector. The levels of trust for the voluntary sector in the entire EU15 are among the highest (after the police and the army). In Britain trust in the voluntary sector is the second highest figure (after the army). Another observation is that the level of trust in the voluntary sector (Fig. 1.5) has also dropped, which can be seen as an indication that trust in the sector is not growing based on the figures of the four Eurobarometer national reports (Reports 58–61). More recent reports did not include the levels of trust in the voluntary sector. However, a more recent study of the Charity Commission[12] indicated that the overall trust mean score was 6.3 (on a 0–10 scale) suggesting "This is a moderate score: although it is not poor, it certainly does not allow for complacency and trust and confidence in charities will require careful monitoring" (Charity 2005).

Another possible reason that contributed to the nonprofit sector being interested in partnering with businesses could be the need for a higher *level of sophistication in funding*. In fact, according to Sargeant (1995) the UK charity sector was going through a transformation as a response to the increasingly 'hostile' environment they had to face in order to survive. That was mainly due to the government allowing even more charities to register, in order to "help subsidise areas that were hitherto supported by the state" (ibid:14) and which inevitably compete for funding. Also a small number of large charities were dominating the market, a trend that continues today. In England and Wales there are 188,000[13] charities growing by about 1,800 per year since 1990 (Strategy Unit 2002). It is interesting that out of a total charity income of £26.71 billion for 2001, 372 charities (representing 0.19% of the total of charities), whose income exceeds £10 million, received more than one third of the income (ibid). The average amount of income these 372 charities receive is almost £24 million per year and less than £1 million for the 187,682 remaining charities.

Consequently, although the amount received by businesses has never been detrimental for the survival of the nonprofit sector, for the largest charities, however, it represents an important amount of funding that derives from a single source (i.e. one company), unlike fundraising from the public where small amounts of money derive from many individuals. According to the NCVO's UK Voluntary Sector Almanak of 2004 (Fig. 1.6), the private sector contributes 4.3% (for 2001/02: £0.86 billion according to the NCVO figures and £0.76 billion according to CAF-Charities Aid Foundation).

Similarly the money received from the profit sector has a symbolic meaning associated with credibility and professionalism as, in general, "success in fundraising

---

[12] The Charity Commission in the UK is the national regulatory body of charities.

[13] The figures presented here are from the Prime Minister's Strategic Unit. It seems that there is no agreement between the different figures presented from the above source and the ones presented from the NCVO (National Council for Voluntary Organisations) neither on the total income of the sector or the total number of charities in the UK. As a result I suggest that the above figures are only indicative of the reality. As an indication the NCVO's figure for the total number of charities in the UK in 2001/02 is 153,000 receiving a total of £20 billion. The same figures from the Strategic Unit are respectively 188,000 (for 2001) receiving £26.71 billion.

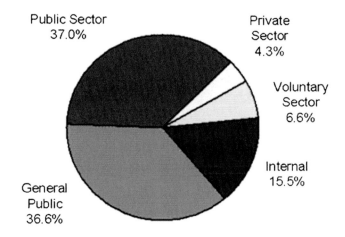

Fig. 1.6 Voluntary sector income 2001/02 (Adopted by NCVO's UK Voluntary Sector Almanack 2004)

links closely to a nonprofit's level of market orientation" (Bennett and Sargeant 2003:799). Furthermore, corporate funding can increase the "brand equity, image and reputation" (Bennett and Sargeant 2003:799) for the business but also for the nonprofit organisation from the association created with a particular 'successful' business. However, with the levels of trust towards the profit and the nonprofit sectors dropping (even if only slightly for the latter) it is important to understand the significance of the positive outcomes that each partner aims for through the partnership relationship as well as the outcomes for society. These questions will be revisited within the chapters that follow.

Finally, the empowerment of business vis-à-vis the rest of the macro forces discussed above contributed to the salience of discussions with regard to the responsibilities of businesses within the globalised social reality. As discussed in the section of 'definition of terms' above, CSR in the UK has been institutionalised (Moon 2004) and hence encouraged by the government. With regard to the meso forces the institutionalisation of CSR prompted collaboration across the sectors in an attempt to offer solutions to social issues that extended beyond the organisational boundaries of the sectors (Seitanidi 2008). Partnership is one of those forms of collaboration across sectors that function both within and beyond organisational boundaries. Hence in the case of the UK, the *institutionalisation of CSR* has been one of the enablers for the promotion of the institution of 'partnership'.

All the above macro and meso forces on the international and national levels contributed to the *generalised partnership phenomenon* in the UK, which does not refer anymore to the sectoral relationship, either within or across sectors, but also to the devolution of responsibilities across all relationships. In fact, 'partnership' has turned out to be a remarkably successful word; hence it appeared on credit cards, logos of organisations, insurance companies, to mention only a few examples. An indication of the popularity of the word partnership in 2000 is given by Jupp

(2000:7) who remarks: "the idea (of partnerships) is so popular that MPs referred to partnerships a staggering 6,197 times in Parliament last year compared to just 38 ten years ago". The denotative power of the word partnership lies in its ability to communicate the message of 'mutual benefits' with just one word which in effect suggests equality between the partner organisations. As it is relatively early for criticism towards partnerships to become a property within the social consciousness and regulation has not caught up yet with the partnership phenomenon[14] we are currently experiencing the 'partnership society'. The 'partnership society' refers to genera-lised fascination with the word 'partnership' by trying to reap its currency of prom-ised mutuality to partners, assuming that this is possible and plausible in all cases. Partnerships for some appear almost as a panacea for all social ills.

The focus of this study is one type of partnerships, i.e. NPO-BUS Partnerships within a developed economy. In the developed economy of the UK the density within and across networks is more intense, NPOs play a more important and widely accepted role both by BUS and government. Civil society is also far more experienced in exercising criticism and more informed due to the level of affluence. Similarly globalisation, the communications revolution and their effects are experi-enced more intensely by all sectors of society. As a result the UK presents a far more complex social order providing a representative example of the Anglo-Saxon model of a developed economy. The uniqueness of the UK case, contributing to the importance of the study but also justifying the selection of this particular country, stems from the fact that partnerships are institutionally encouraged by the government. Furthermore, examining partnership within a developed economy is important, because it allows for the examination of each sector's influence within the partnership relationship e.g. forming equally the partnership agendas and the extent to which they are both able to assume their role as emerging powerful actors as suggested previously.

The partnerships between NPO-BUS organisations are embedded into a network of relationships across all sectors of society and institutions. The partnership programmes across big organisations are seen as an expression of the corporate and nonprofit cultures, their understanding of what constitutes a social issue and the priorities that need to be addressed. Hence their decisions shape not only the solutions of the problems they address but also, indirectly, partnerships have a broader political significance. Partnerships, as a new way for societies to govern themselves, allow organisations from different economic sectors to jointly prioritise social issues, hence shaping social priorities. Their decisions have an impact on a network of social institutions and organisations, hence the power of partnerships to be agents of social change. Himmelstein (1997) in his book about corporate power and corporate philanthropy in the US 10 years ago, presented the potency of the corporate philanthropy network culture. He argues that corporate giving as a complex

---

[14]I refer here to cross sector partnerships rather than civil partnerships or limited liability partnerships as the latter two are regulated forms of interaction, the first within family law and the second within company law.

social and political act becomes politicised regardless of the claims of big corporations that their work is apolitical and non-ideological. In fact he suggested that "patterns of giving are often reinforced by powerful, durable norms" (ibid 147). The philanthropic agendas of the affluent are guided by their interests and priorities; in effect it is unlikely that they would be changing their agendas in order to take over the tasks that were previously implemented by the government "because their priorities are shaped by a strong culture of philanthropy among the social elite" (Himmelstein 1997:147). This type of decision is supported by a powerful network comprised of members of elites (business colleagues, trustees of NPOs, members of clubs, relatives) that reinforce and perpetuate the network culture, its priorities and ultimately form decisions. NPO-BUS partnerships, as successors of corporate philanthropy, appear to be a widespread phenomenon both in the UK and globally. This book examines within the partnership society case studies on NPO-BUS partnerships and how they emerged within the context of the UK.

1.5   Conclusion

The first chapter of the book clarified the type of social partnership under investigation within this study: NPO-BUS partnerships, a type of social partnership between the profit and nonprofit sectors. It discussed briefly the responsibilities and roles of the dominant model of the tri-sectoral divide of society in order to highlight the changes that are occurring. It further discussed the definitions provided by the literature on 'nonprofit organisations, NGOs, civil society organisations' and 'social partnerships' in order to establish a platform of common understanding for the terminology employed in the book. It also presented the historical/organisational context of collaborative relationships between businesses and nonprofit organisations by examining the collaboration continuum framework and discussed the macro and meso forces that contributed to the generalised partnership phenomenon, or in other words the 'partnership society'.

The chapter put forward the elasticity of the sectoral boundaries that society is experiencing today concerning the roles of the sectors, through the discussion about the delineation of roles and responsibilities within and across sectors. It maintained that partnerships are addressing issues beyond the traditional organisational boundaries based on the discussion of the definitions of social partnerships. Hence, NPO-BUS partnerships, as one of the types of social partnerships, lie within the social arena that was traditionally termed 'public policy'.

Furthermore the chapter discussed the macro forces that relate to the global phenomena that impact on the relationships of the sectors, namely globalisation, the lack of international legislation, the decline of the old role of the state, the empowerment of business, the communications revolution, the empowerment of civil society organisations and the changes in consumerism. It moved on to discuss the meso forces that occur on the institutional level in the UK, which comprises all three sectors, and refer to the sectoral dynamics. The meso forces that contributed to the partnership phenomenon between the profit and nonprofit sectors in the UK are: the national devolution of power, the fragmentation of delivery, the crisis of legitimacy, the deficit of governance, the decline on the levels of trust, the increased levels of sophistication in funding and finally the high level of institutionalisation of CSR in the UK.

The above macro and meso forces on the global and national-sectoral or institutional levels contributed to the generalised partnership phenomenon in the UK, the meaning of which extends beyond the level of relationships referring to the devolution of responsibilities across all sectors.

The focus of this study is the NPO-BUS partnerships within a developed economy, i.e. the UK. The importance of studying these types of relationships within a developed economy can provide an indication of the challenges the existing sectors are facing today and can show how they perform within the context of a partnership in a developed economy where such relationships are institutionally encouraged. It can offer an insight into the reasons of the organisational context that set in motion the boundaries of the sectors and the implications for either the organisational actors or society at large.

Partnerships have broader political significance. The large organisations involved in partnerships, both businesses and nonprofit organisations, have the power to form associational trends and funding patterns that are followed gradually by other organisations. Also as the associational 'tone' is set in the early stages by a few high profile organisations; politically it offers the justification for others to proceed with similar relationships. Hence their decisions shape not only the solutions to the problems addressed, but also the sectors these organisations represent. Today corporate power is partnering with nonprofit credibility hence the politics in the micro-associational domain are important for the credibility and accountability of the role of the sectors and ultimately the changes that will emerge, which will determine the extent to which social partnerships will develop sustainable solutions for society. The next chapter discusses the theoretical constructs that are deemed important within the micro-associational domain of a partnership relationship.

Chapter 2
A Framework for the Analysis of Partnerships

2.1 Introduction

The second chapter situates the current study within a theory-based framework that is put forward for the analysis of NPO-BUS partnerships. The chapter proposes exploring questions in each of the three stages of partnerships: (a) formation, (b) implementation, and (c) outcomes by studying them within the same partnership case study. Employing the three stages of partnerships and associating each stage with particular constructs allows for a holistic and in-depth examination of partnerships. It also allows the analysis to move beyond the stages and stage-based questions towards identifying overarching themes that encompass all stages. The chapter, which is grounded within organisation theory and more specifically the NPO-BUS partnership literature, addresses theoretical issues that play a central role in NPO-BUS partnerships, while at the same time reviews the literature by presenting the findings of indicative studies. The chapter commences with an introduction on the relations between nonprofit organisations and businesses, followed by presenting a classification of the main contributions of the literature. The remaining three sections present the assumptions and questions that this study addresses within its empirical chapters grouped under the three stages of partnerships.

2.2 Relations Between Nonprofit Organisations and Businesses

Socially responsible practices of businesses refer to the entire range of policies and practices of a company that aim at treating their internal stakeholders (employees, suppliers, shareholders) and external stakeholders (communities, customers) ethically, with integrity, while respecting both basic human rights and sustaining the environment in which they operate. The social responsibilities touch upon all business

M.M. Seitanidi, *The Politics of Partnerships: A Critical Examination*
of Nonprofit-Business Partnerships, DOI 10.1007/978-90-481-8547-4_2,
© Springer Science+Business Media B.V. 2010

functions such as procurement, production, marketing, human resources, external relations, or as it is most often called today, community involvement, to mention only a few. The interest of this research concentrates on the last type of function, i.e. corporate community involvement, and in particular its sub-function of improving the corporation's relations with communities by developing relations with NPOs. There are different forms of collaboration with nonprofit organisations, one of which is partnership,[1] the focus of this study.

The last 200 years have witnessed the intensification of the interaction between the private and nonprofit sectors (Galaskiewicz and Colman 2006; Gray 1989; Young 1999; Austin 2000; Googins and Rochlin 2000). These interactions occurred within the broad practice area of 'corporate philanthropy/support' (Wymer and Samu 2003; Wragg 1994; Kotten 1997; Meenaghan 1984) later replaced by the more general term of 'corporate community involvement' (Brammer and Millington 2004; Moore 1995; Patterson 2004) also referred to as 'community relations' (Waddock 2001). During this time a number of differing forms of interaction between these two sectors have emerged, including: corporate philanthropy (Kotten 1997); sponsorship (Meenaghan 1983) and cause-related marketing (Berglind and Nakata 2005; Varadarajan and Menon 1988). However, the gradual prominence of the concept of corporate social responsibility (CSR) within all sectors of society elicited an intensification of the debate with regard to the responsibilities of each sector in addressing environmental and social issues. In effect CSR contributed to the increase of the interactions across the sectors and propelled NPO-BUS partnerships (a type of social partnership) as a key mechanism for corporations to delve into a process of engaging with NPOs in order to improve their business practices by contributing their resources to address social issues (Heap 1998; Mohiddin 1998; Fowler 2000; Googins and Rochlin 2000; Mancuso Brehm 2001; Hemphill and Vonortas 2003).

The literature on relationships between NPOs and BUSs reflects the shift of the industries it studies. While in the past the social responsibilities of businesses were manifested mainly through philanthropic relationships, in later years (after 1997) the form of partnership was increasingly the new terrain that academics in this area of study have been exploring. The papers published in the mid-1990s were mainly from scholars in the USA (Stafford and Hartman 1996; Hartman and Stafford 1997; Waddock 1988) with a few exceptions only towards the end of the 1990s as the UK has followed after the first indigenous partnerships appeared in the country (Murphy and Bendell 1999; Crane 1998).

In capitalist societies, businesses are seen as one of the primary causes of environmental and social problems. The partnership literature (Waddock 1988; Bendell 2000c; Austin 2000) primarily advocates that businesses should equally

---

[1] Other forms of collaboration include sponsorship, socio-sponsorship, cause-related marketing, philanthropy, benefaction and patronage. For a comparative review of the different forms of collaboration see Seitanidi and Ryan 2007.

participate in offering solutions to environmental and social problems and in effect change their practices in order to address their responsibilities. Three propositions exist with regard to the role of the nonprofit sector in the process of achieving change as one of the outcomes of partnerships. The first proposition suggests that change can be achieved by fostering a confrontational approach towards businesses, i.e. exercise pressure towards a BUS without forming a collaboration (Bendell 2000b); the second suggests that the collaborative approach can be more effective as it allows NPOs to change BUSs from within (Bendell 2000b; Heap 2000; Bray 2000). The third or the 'middle way' of 'critical co-operation' (Covey and Brown 2001) suggests that by having a collaborative approach, confrontation and collaboration can take place within the relationship. However, the context of the normative conception of 'critical cooperation' was rooted within the previous assumption of an apriori conflict that existed between the business and nonprofit organisations. So the starting point of the type of interaction suggested by Covey and Brown (2001) is different from the one presented in this study, which in fact questions this fundamental assumption. This study suggests that partnerships between NPOs and BUSs belong to the collaborative approach of association as opposed to confrontational approaches (Bendell 2000c; Crane and Matten 2004) and that we have moved to a different reality across the sectors on the organisational level, which presents a new context in which theoretical or empirical questions need to be asked.

According to a number of scholars the relationship between NPOs and the corporate world has increased, intensified and became more important (Galaskiewicz and Colman 2006; Wymer and Samu 2003; Stafford and Hartman 2001; Googins and Rochlin 2000; Waddell 2000; Kanter 1999; Murphy and Bendell 1999). Although the above could be the case, there are no studies that compare the earlier type of relationships between the two sectors such as socio-sponsorship or cause-related marketing (CRM) or even earlier corporate philanthropy in order to offer comparative evidence of the increase. In fact, based on what partnership is about and what it involves we should have witnessed a decrease in the number of relationships between the two sectors. As partnerships are resource intensive, one of their characteristics is that a larger sum of money is given from a company to fewer nonprofit organisations hence the total number of relationships should have decreased. In effect the assumption of the increase in relationships is yet under-explored in the literature due to the lack of comparative quantitative studies across different forms of association.

The types of partnerships that the book explores concentrate primarily on the alignment of strategic business interests with societal expectations (Covey and Brown 2001; Austin 2000) expressed through NPOs. The aim is to understand key processes in the development of a NPO-BUS partnership, particularly when the partners appear to have a high degree of compatibility among them or in other words when they do not have conflictual interests or previous conflicting relationships from the outset. The next section aims to position the study within the NPO-BUS partnership literature.

2.3   The Study of NPO-BUS Partnerships Within the Literature

Although the literature on NPO-BUS partnerships discusses theoretical issues, it appears, mainly due to the fact that it is relevantly a recent phenomenon, that most studies do not engage in placing partnerships within a single theoretical context, but rather encompass a broad array of disciplines such as "organisation studies, public policy and administration, economics, nonprofit management, health care, education and the natural environment" (Selsky and Parker 2005:2). Table 2.1 offers indicative examples of studies on NPO-BUS partnerships with the main objective to identify groups of interests within the NPO-BUS partnerships literature by categorising a number of studies and their main contributions. The table describes the different strands identified, with a summary of each category's characteristics and representative studies with the name(s) of the author(s) and the year of publication. Although there are studies that contribute in more than one of the areas identified, this categorisation considered the main contributions of each study.

Six different strands are identified within the literature of NPO-BUS partnerships:

1. *Nature of Partnership*: These studies are primarily concerned with analysing or providing a definition of NPO-BUS partnerships either in a national context or cross-nationally. They identify the characteristics of the partnership phenomenon within a specific industry or country under study.
2. *Managerial Aspects of Partnership*: The second strand is characterised by the managerial aspects of partnerships, such as the objectives of the profit or the nonprofit organisation, the motivations, the organisational structure of each partner involved or of the industries, and either the personnel or volunteering aspects of partnerships. A number of studies also describe the process of partnership development.
3. *Strategic Use of Partnership*: This category of studies is mostly concerned with teaching the readers of the study how to build successful partnerships. The main focus of the authors is to identify the strategies that worked well in the partnerships under study and convey messages for the strategic use of the partnership, or in other words to develop partnerships for achieving sustainable development. These studies are rarely concerned with offering generalisations outside the industries under study.
4. *Legal and Ethical Considerations in Partnerships*: The issues of accountability within the partnership, legitimacy and transparency are usually the focus of these type of studies. Also culture and power dynamics are discussed in order to assess the contributions of the partners involved.
5. *Partnership Measurements*: This strand would involve figures around partnership such as comparisons across different partnerships, for example, environmental partnerships in order to assess the financial contributions, the total costs involved within a partnership or the amount of hours invested. Furthermore providing estimation for the direct or indirect audiences or target groups that were

Table 2.1 NPO-BUS partnership strands within the literature

Partnership strand	Representative studies
1. **Nature of partnership:** Defining partnership, identifying characteristics. Partnership evolution in a particular country or industry	Wymer and Samu (2003), Covey and Brown (2001), Elkington and Fennell (2000), Murphy and Coleman (2000), Schneidewind and Petersen (2000), Austin (2000), Heap (2000), Murphy and Bendell (1999), Bendell (1998), Stafford and Hartman (1998), Murphy and Bendell (1997)
2. **Managerial aspects of partnership:** Objectives, motivations, organisational structure, personnel requirements, volunteering, budgeting, process of partnership development	Backer (2007), Andrioff (2000), Austin (2000), Fowler and Heap (2000), Crane (1998)
3. **Strategic use of partnership:** Identifying strategies associated with partnership, suggesting 'how to'	Warner and Sullivan (2004), Loza (2004), Bendell and Murphy (2000), Turcotte (2000), Waddell (2000), Moser (2001), Nelson and Zadek (2000), Stafford and Hartman (2000), Waddock (1988), Stafford and Hartman (2001)
4. **Legal and ethical considerations in partnerships:** Issues of accountability, legitimacy, transparency, cultural and power dynamics	Parker and Selsky (2004), Hardis (2003), Tully (2004), Millar et al. (2004), Crane (2000), Bendell and Lake (2000), Bendell (2000a)
5. **Partnership measurements:** Partnership figures, monitoring, estimating direct or indirect audiences	N/A
6. **Societal implications:** Drawing societal inferences based on the study of category (1) (nature of partnership) and (2) (managerial aspects of partnership) above	Hamman and Acutt (2003), Millar et al. 2004, Tully (2004), Seitanidi (2008)

reached through partnerships is another category that could be included in this strand. Based on the literature review there are no studies that can be included in this category or indeed no estimation of the total turnover that the institution of partnership is able to evoke.

6. *Societal Implications*: Finally, this strand primarily draws links between the nature of partnership and the managerial aspects of partnership with the aim of eliciting and discussing the societal implications as a result of the NPO-BUS partnerships. Hence, identifying the characteristics of the partnership phenomenon within a specific country or examining the partner's motives in order to formulate inferences on the societal level.

In their recent review[2] of cross-sector partnerships Selsky and Parker (2005) suggest the emergence of a new platform for cross-sector partnerships: 'the societal sector platform'. The main thrust of this platform is that the relationships across the sectors are blurring the boundaries between the traditional roles of the sectors, hence one sector "adopts or captures a role or function traditionally associated with another sector" (ibid:5). The above identified strand 'societal implications' suggested by this research draws attention to the outcomes of these relations, hence it is different from the societal platform suggested by Selsky and Parker.

As it becomes apparent from the above table, the categories that provoke more interest among academics become evident. The 'study of the nature of partnerships' and the 'strategic use of partnership' draw the majority of attention from academics. This is justifiable based on the fact that the study of NPO-BUS partnerships is a relevantly new phenomenon for academics, and since it is in its early stages its 'nature' needs to be examined, hence the focus of attention of researchers. With regard to the 'strategic use of partnership' category, it equally responds to the demand of the 'promising' mutual benefit relationship that partnership can deliver; the managers across the different sectors are at the receiving end of this category.

The findings of the above NPO-BUS literature are presented below to offer an indication of the issues discussed. Within the 'nature of partnership' Covey and Brown (2001) discuss intersectoral co-operation by describing NPO-BUS initiatives, suggesting the application of the concept of critical co-operation as a possible framework that can result in the productive engagement of both sectors. The paper offers a theoretical contribution based on a discussion of the rights and responsibilities of the sectors. However, it does not examine if the power dynamics of the partners allow for the conflict to externalise within the partnership relationship. Elkington and Fennell (2000) discuss NPO-BUS partnerships as an emergent phenomenon, examining the reasons that brought the two sectors together. They propose a typology of NGOs assisting in the understanding of the nonprofit sector but also offering a reflection of the NGO campaigners and their strategies towards business. The suggested typology of NGOs reflects the dichotomy of the sector between pro-business and radical NGOs suggesting that perhaps 'critical co-operation'

---

[2] Their review extends to all types of social partnership, not just NPO-BUS partnerships.

could be applied only within the pro-business NGOs, if at all. Murphy and Coleman (2000) discuss the NPO-BUS partnership phenomenon as a 'mutual symbiosis' between the sectors and partners, suggesting that partnerships have the potential of changing the way businesses work and ultimately of transforming society without indicating how such a major social change would be possible. Schneidewind and Petersen (2000) apply structuration theory to three short case studies within the German context in order to assist the understanding of business collaborations with environmental nonprofit organisations suggesting how companies can become responsible actors in favour of sustainability in social-structure building. Their examination of secondary data puts forward the need for further empirical research in the field with more attention to detail on processes. Austin's (2000) work has been summarised in the previous chapter, as his main contribution is the collaboration continuum framework drawing the characteristics of each stage of collaboration allowing for comparisons to be drawn across the stages. Heap's (2000) book on relationships about NGOs and businesses is an informative introduction to a wide variety of issues that encompass these relationships. One of the findings is that these relationships in the case of environmental partnerships are seen as "an engine for change" (ibid:260); also that ENGOs (environmental NGOs) are a decade ahead of the DNGOs (Development NGOs) in terms of organisational capacity and outlook. Heap also suggests that there are no differences in the generic engagement rules between the "North and the developing world of the South and East … as engagement appears to be issue-specific" (ibid:259). The last finding is shared by Andrioff's study, which is discussed below. Murphy and Bendell's (1999) paper offers "a global overview of the changing nature of business-NGO relations on sustainable development" (ibid:1) highlighting the value, the impact, the politics and processes of such relationships. They offer the term of 'civil regulation' as an important driver for corporate and environmental responsibility. Bendell (1998) introduces the term of 'civil compliance' whereby companies abide by society's demands raised by 'civilians', expressing the active involvement of individuals in the wider societal issues. Murphy and Bendell (1997) offer an early exploration of NPO-BUS relations with an emphasis on environmental partnerships for sustainable development providing a critical review of a number of examples. Stafford and Hartman's (1998) paper offers a typology of environmentalist-business co-operation and "argues that a sequence of external-internal political economy forces are driving the most complex co-operative forms" (ibid:62) calling for future research to address the consequences of environmental-business co-operation.

In the second strand, Backer (2007) argues by presenting a case study on Shell that the environmental decision making in business can change through the diffusion of corporations' environmental governance practices as they are being impacted through their interactions with NPOs; Andrioff's (2000) research provides an insight into the implementation of social risk management through stakeholder partnership building from a business perspective. It offers a comparison across the case studies, identifying a number of characteristics in partnership building according to their purpose. He concludes that the partnerships were partnership-specific rather than context driven. Crane (1998) presents green alliances through an exploratory case study discussing the alliance motives and the inter- and intra-organisational relationships and cultures. He identified congruity of bonds between the partners, resources, activities and cultural mediation as critical factors for the success of partnerships. In addition, Fowler and Heap (2000) analyse the motivations, organisational issues, key benefits and challenges within the case study of the Marine Stewardship Council.

In the third strand Warner and Sullivan's (2004) book explores tripartite partnerships within the development context, primarily discussing how partnerships work and how to make these relationships more effective in a wide variety of national contexts. They suggest that partnerships "enable communities to take charge of their own development needs, interacting with government to jointly design and maintain public services" (ibid:345). Loza (2004) focuses on the benefits for community organisations when partnerships are aiming at building organisational capacity for NPOs and the benefits for businesses as well. Turcotte (2000) showed in her analysis how an environmental NGO "through its different type of relationships with business organisations, helped shape aspects of the economy in order to make it less environmentally damaging" (ibid:134). Waddell (2000) presents from a business perspective what partnership can bring to core corporate functions "driven by 'win-win' or by a 'mutual gain' perspective". Moser (2001) focuses on MNC's (multinational corporations) contribution to sustainable development of less developed countries proposing the concept of sustainable business practices. One of his findings is "the importance of external institutional pressures – principally in the form of local legislation – for determining the contribution of MNCs to the sustainable development of less developed countries" (ibid:305).

Nelson and Zadek (2000) in their cross-European report on the new social partnerships in Europe present a very wide range of managerial issues that take place in partnerships primarily promoting learning about the phenomenon. One of their conclusions is that "a major shift is taking place in our understanding and practice of governance.... Governance is today increasingly about the roles, responsibilities, accountabilities and capabilities of different levels of government (local, national, regional and global) and actors or sectors in society (public, business and civil society organisations)" (ibid:55). They remark that both the nonprofit as well as the profit sectors participate more actively in the evolution of public policy and delivery. They claim that this is a result of "greater interaction, trust and intimacy between these two groups and in part due to the shifts in conditionality of funding ..." (ibid). However, they do not provide clear justification about the above assumptions and more importantly of the possible negative implications of these changes. It hence remains unexplained the increase of more radical organisations today in the form of network associations that do not use the traditional structure or systems of organising. Therefore, the increase in trust that is claimed above does not really represent the reality we are experiencing today of diminishing levels of trust as discussed in the previous chapter. Stafford and Hartman (1998) address the important issue of credibility of environmental partnerships. They maintain that the social acceptability of these relationships "is likely to be a concern among consumers and other stakeholders" (ibid:189). Although they suggest a framework that marketers can employ to advance the relationship and organisational credibility, they do not address the negative implications to stakeholders when these relationships are primarily benefiting the two partners rather than society at large. Waddock's (1988) paper on building successful partnerships provides one of the most cited definitions of social partnerships (discussed in the previous chapter). The paper refers to the dangers inherent in partnerships, but primarily suggests how to build a successful partnership by understanding the limits of such forms of association. Stafford and Hartman's (2001) paper discusses the diffusion of environmental technology innovation around the world by analysing the 'Greenfreeze' campaign of Greenpeace. They define 'creative destruction' as "the revolutionary changes that simultaneously destroy the established economic system from within and create new markets, industries and organisational relationships. … Creative destruction is inherently threatening to incumbent firms who are likely to resist radical technological changes and act in ways to preserve their market positions and profits" (ibid:107). They identify a number of factors that can be employed by other environmental technologies in order to market them successfully. It is very interesting that in their compilation of factors they examine the partnership as a whole, including different level of interactions across the partners, the industries involved and the broad implications of strategies.

In the fourth strand of the literature, Parker and Selsky (2004) move the discussion of 'cause-based partnerships' from the a priori differences approach that characterised the majority of the literature (in demographic, task, process and culture) to an emergent culture approach. They emphasise that the emergent meaning in partnerships is interactively evolving along with the power dynamics and trust among the partners. Hardis (2003) primarily presents the process of collaboration in the case of social multipartite partnerships and concludes "the practice does not follow the rhetoric" (ibid:208). She reiterates the need for more empirical research in a different way "that would take into account more dynamic, fragmented ways, and institutionally defined aspects of partnership dialogue and meaning construction" (ibid:209). Tully (2004) assesses the prospects of 'civil' regulation and the ability of NPOs to regulate the behaviour of corporations within the frame of partnership. He concludes that partnerships as a form of participatory regulation are dependent upon: (1) the legitimacy of the arrangement, (2) the accountability of the partners, and (3) the nature of the decision-making process. Furthermore, Tully remarks that NGOs face higher reputational risks than businesses which enjoy higher reputational benefits in a partnership relationship. He suggests that as partnerships can be seen to encourage de-regulation it is important for governments to intervene in order to facilitate their effectiveness, since at the moment the partnership outcomes are reliant on self-regulation. Millar et al. (2004) provide a conceptual framework highlighting the importance of the dual roles of NGOs based on their market and institutional identities. They raise a number of ethical issues within the context of NPO-BUS partnerships while referring to criticism for both sectors. Finally, they call for further research concentrating on the role of NGOs in the mature, developed economies, the emerging economies and the economies in transition. Crane (2000) explores NPO-BUS collaborations from an organisational culture perspective examining the possibilities and problems of 'culture clash' between these diverse partners. He suggests that cultural disharmony can be addressed by 'cultural mediators' (certain individuals or organisations within the alliance) that can act as bridges by establishing common meaning and understandings. Bendell and Lake (2000) discuss the new frontiers between the two sectors in transparency, accountability and financing. They maintain that the power shift that takes place in society will further push the frontiers to widen and deepen supporting the agendas of corporate social responsibility and sustainable development. Bendell (2000a) challenges the 'win-win' scenario that predominates in the literature and examines the 'win-lose' scenario within the context of biotechnology companies. He suggests that biotechnology companies could work with CSOs in order to include them in their decision-making processes. This is the only paper that challenges the 'win-win' concept associated with partnerships, although only within the biotechnology industry.

Finally, in the sixth strand Hamman and Acutt (2003) provide a critical examination of CSR (Corporate Social Responsibility) and NPO-BUS partnerships. They refer to critical co-operation as a possible alternative to these types of relationships, suggesting that improving BATNA (Best Alternative to a Negotiated Agreement) will assist NPOs to consider a priori not only the interests and rights issues but power as well. In other words, NPOs need to exercise their power before and during their partnerships with a business in order to play an active role in shaping CSR discourse. In this category the papers of Tully (2004) and Millar et al. (2004) can also be included as they draw inferences for society at large, even if their major contributions are classified within the fourth strand of the literature. Finally, Seitanidi (2008) presents an instance of failed large scale social innovation from a cross sector social partnership even though the partnership seemed to succeed in its narrow mission. The paper is calling to move beyond reactive and proactive responsibilities and to shift towards accepting adaptive responsibilities that require a multidimensional understanding towards all three levels of analysis, micro, meso and macro in order to externalise the benefits of partnerships to the societal level.

Based on the above review of the literature it appears that the majority of the research on NPO-BUS partnerships has a strong focus on the instrumental orientation (Selsky and Parker 2005) that follows firstly from the practice of partnerships and secondly "reflects the resource dependency argument in organization studies" (Selsky and Parker 2005:10). The more recent critical papers appear in the last category and indicate entering a mature phase in the examination of the phenomenon.

Furthermore, within the above strands we can observe further sub-categories:

1. *Perspective Adopted*: A single actor perspective, dual or a societal perspective employed. In the first category the researcher adopts one perspective to look at the study: either the profit or the nonprofit sector's perspective, for example: Warner and Sullivan (2004); Austin (2000) discuss partnerships primarily from the business perspective. In the second category the study adopts a dual perspective trying to show how each participating partner, both the profit and the non profit organisation, benefited from the relationship. In the third category the

study includes society as well as the participating partners in order to question the extent to which society at large benefited from the partnership.

2. *Type of Analysis Adopted*: Predominately the majority of studies do not criticize the partnership phenomenon. An explanation of that could be that perhaps the majority of the studies adopt a more positivistic epistemological position, and as a result accept the reality of the participating organisational actors as the single existing reality. The organisational actors on the other hand perceive or portray their partnerships as successful in their effort to justify their realities. However, there are studies that do not follow this tradition and which contribute a more critical approach to the literature such as Tully (2004), Millar et al (2004), Crane (2000), and Bendell (2000a).

The current study belongs to the sixth strand as it applies a societal sector approach. It discusses the societal implications of NPO-BUS partnerships by studying both the nature of NPO-BUS partnerships in the UK and the managerial aspects within the partnerships under examination aiming to make inferences on the institutional level of the sectors represented within the partnerships under study. It further adopts a critical perspective towards both partners involved and hence their respective sectors. The research questions that the book is addressing are presented below under the sequential stages of partnerships which are used as a framework to present both the literature and later the findings.

2.4 NPO-BUS Partnerships: Formation-Implementation-Outcomes

According to the paper of Selsky and Parker (2005:6) "researchers almost universally agree CSSPs[3] can be examined according to chronological stages". The literature does not offer one model stage but rather authors have suggested different models (Googins and Rochlin 2000; Gray 1989; Waddell and Brown 1997; Westley and Vredenburg 1997). The book examines the stages of partnership building, which are referred to as 'phases'; it employs the phases of formation, implementation, outcomes in order to group firstly the theoretical issues examined in the research and secondly to group and present the findings in the empirical chapters. As Selsky and Parker (2005) report in their review paper different authors concentrate on different stages of the partnership relationship. This study examines all three phases within two in-depth case studies in order to allow for observations that encompass the three phases and hence arrive at conclusions about the type of organisations involved and the implications at the societal level. It also supplements the cases with comparative interviews that allow for further confirmation or disconfirmation of the findings within a wide range of organisations.

---

[3] CSSP refers to cross-sector partnerships that address social issues.

The three phases and the constructs examined under each phase are: (1) formation, under which the organisational characteristics of the partners are examined, the historical evolution of the relationship and the motives that are associated with each partner; (2) implementation, examines the process or phases of partnership building and the dynamics between the partners; (3) outcomes which concentrate on the organisational, social and societal benefits.

2.4.1 Partnership Formation

Forming a partnership is a process that starts before the existence of a partnership relationship, proceeds through the early stages, when the two organisations firstly develop the partnership, until the maturity stage and thereafter.

In contrast to the traditional business-to-business relationships, cross-sector collaborations are usually characterised in the literature as 'non-traditional alliances' due to their "complexity" (Kanter 1999:126) and also because of the inherent "contradictions and conflicts between incompatible objectives, ideas and values" (Holzer 2001:9). The differences between same-sector collaborations and cross-sector collaborations include: "different performance measures, competitive dynamics, organisational structures, decision-making styles, personal competencies, professional languages, incentive and motivational structures and emotional content" (Austin 2000:14). As Kanter (1999:126) points out, nonprofit organisations are driven by goals other than "profitability and they may even be suspicious of business motivations". McFarlan (1999) suggests that the goals and characteristics of nonprofits are different from those of for-profit organisations concerning governance. Similarly, their values, motives and types of clients are different (Di Maggio and Anheier 1990); their objectives are sometimes conflicting, especially in the case of environmental alliances (Stafford and Hartman 2000).

As this suggests, the literature on cross-sector alliances portrays the partners from different sectors as different and distinct from each other with regard to their missions. It further suggests that this tends to create a degree of conflict among the partners (Shaffer and Hillman 2000; Westley and Vredenburg 1997) and distrust (Rondinelli and London 2003). In particular in the context of environmental partnerships misunderstanding of each other's motivations (Long and Arnold 1995) often undermines the formation and implementation of these relationships (Rondinelli and London 2003). Hence the majority of the literature fosters a strategic perspective (Birch 2003; Jupp 2000; Hartman et al 1999) that aims to present how these relationships can be managed successfully. These studies present the skills that are needed to pursue successful cross-sector collaborations. They further claim the historical adversarial origins between business and nonprofit organisations are pronounced particularly in the case of environmental NPOs (Rondinelli and London 2003). Another aspect that the literature concentrated upon was the unfamiliarity among the partners (Rondinelli and London 2003) which resulted in considerations for each partner with regard to the influence of these relationships upon their reputation and public image.

Although the literature on NPO-BUS partnerships makes strong claims about the pre-existing differences and conflicting missions between the two partners, for the most part these claims reflect the situation at the early stages of the phenomenon. Today a number of factors resulted in the initiation of recent changes within the institutional environment that are already observable in the fine detail. The increased interaction among the sectors, the increase of registered charities in the UK (Sargeant 1995) and the decrease of government support led to the intensification of the competition within the nonprofit sector for funding (ibid). Furthermore, the empowerment of corporations as important financial and political actors (Newell 2000) in conjunction with the need of NPOs to replace their traditional sources of funding has increased their drive to become more businesslike (Dees 1998). Within the changing institutional environment it is important to question past claims that were only valid in some cases (for example in environmental alliances e.g. Stafford and Hartman 2001). Furthermore, it is also important to examine if the characteristics of NPOs and BUSs are contributing in bringing the two partners together and if indeed the assumed conflict between the partners produces outcomes that are consistent with the rhetoric of partnerships delivering societal benefits.

This study is questioning the broad assumptions of an a priori difference and conflict among the two sectors based on the characteristics of the two partners in the case of a partnership relationship. It aims to investigate the characteristics of the for-profit and NPOs when they engage in a collaborative relationship that benefits society as a result. For this reason a clear distinction is made between social and societal outcomes, discussed below under the section partnership outcomes. The first research question addressed by this study is:

RQ1: What are the organisational characteristics of the NPO and BUS that decide to form a partnership?

With regard to the organisational characteristics in the context of NPO-BUS partnerships, Berger et al. (2004:76–81) examined separately the nonprofit and the company structural characteristics that had an impact on the partnership relationship. For the NPOs they identified five characteristics: (a) programmatic versus grant-making NPOs (the former is the typical charity organising and running programmes for its constituents and the latter raises money to donate to other NPOs); (b) autonomy versus control (NPOs whose regional offices are largely independent of the headquarters as opposed to NPOs whose headquarters have control over the regional offices); (c) big, well-established versus small, entrepreneurial; (d) revenue-generating products (or services) versus none revenue-generating; (e) inherent cross-sector collaborations versus traditional (in the first case NPOs have inherent in their missions to work with BUSs as opposed to traditional NPOs for which collaborating with BUSs remains one of the tactics for generating income). In this last category Berger et al. (2004:81) mention that "as nonprofits begin to mirror business structures and approaches, opportunities for companies to learn from the traditional strengths of nonprofits diminish". The above characteristics, based on the research by Berger et al, have both pros and cons depending on the focus of the partnership and the BUS partner. For example, they suggest different characteristics can be advantageous, offering greater opportunities for BUS employees to get involved, a greater or lesser possibility for innovation, higher or lower 'brand equity', opportunities or constraints for additional revenue, fewer or more cultural barriers to partnering, more or less possibility for exclusivity.

Similarly, the structural characteristics for BUSs, according to Berger et al. (2004:81–83), also have pros and cons, and some of their influences can determine the potential advantages depending on the cause of the NPO offering higher visibility, credibility, allowing for faster or slower decision making, having more or less available funding, increasing the commitment of business to the partnership but also increasing the communication challenges within the relationship. The characteristics they propose are: (a) flat versus hierarchical organisational structures; (b) broad consumer market versus specific target markets; (c) direct sales force or retail presence versus business-to-business; (d) pre-eminent versus less-eminent brands. As they suggest, the above are only propositions that demonstrate a tendency based on their research findings. Since not all of the above characteristics are relevant to this research the empirical and the discussion chapters will further discuss the characteristics that are consistent with the findings of this research.

The historical perspective of the relationship is another important parameter as it both informs the partnership relationship and determines the decision of both partners to proceed to the formation of a partnership. Rondinelli and London (2003:71) remarked that "absent prior relationships or a series of other transactions, as in the case of most cross-sector alliances, potential participants can only rely on the 'shadow of the future' or build contacts that provide for protection against opportunism". Following on from the above and in order to examine the context in which a partnership relationship takes place the study aims to place the relationship in a historical perspective in order to trace the previous interactions among the partners. The previous relationship among the partners plays an important role in informing their decision to develop a partnership relationship. Hence the second research question of the study is:

RQ2: How does the relationship between an NPO and a BUS evolve into a partnership?

The decision of organisations to develop a partnership relationship is guided by their motives, indicating the expected outcomes from the relationship. Mills (1940) argues "that motives are terms social actors use to interpret their present, and guide their future" (cited in Peters 2004). Following Peters (2004), motives are viewed as a way to explain and reinforce actions providing a link between the individual participants and institutionalised situations: "when a vocabulary of motives appeals to a broad base of people in an act or activity, the vocabularies are strategies of action" (ibid:211).

According to Mills (1940), institutionalised situations within a certain epoch and context employ unique vocabularies of motives. Social actors internalise these vocabularies, which in turn justify and validate their normative behaviour. Hence, motives link participants to situations, explain and reinforce acts. Consequently, analysing the partners' motives offers an examination of the organisations' justification for the collaboration and the deployment of their resources for the relationship.

Also the perceptions of motives are closely linked with each organisation's mission and characteristics allowing for observing the type of organisations and in effect the outcomes achieved as perceived by the organisational actors.

Within a partnership, there are two sets of motives in operation: the motives of the profit organisation and those of the nonprofit organisation. Each organisation represents a different system of values and beliefs (Crane 1998; Stafford and Hartman 2000) due to their sectoral characteristics. Corning (2003:1) suggested that "biological survival and reproduction remains the fundamental, ongoing, inescapable challenge for all living organisms, including humankind; it is a problem that can never be permanently 'solved'". Transferring his statement to the organisational level, equally the survival of organisations is the most fundamental concern for either a business organisation or a nonprofit organisation. In the case of nonprofit organisations their existence is justified by serving social needs and stakeholder interests (Bryson et al. 2001; Oster 1995; Salamon 1992). However, for a nonprofit organisation to serve its mission it needs to survive and grow (Bryson et al. 2001). For the profit sector survival is usually expressed as 'sustainability' and in the case of the nonprofit sector as 'organisational survival'. However, both motives in essence refer to the same thing, as they aim to maintain the existence of the organisation. In the case of sustainability, survival does not include only one organisation but at the same time it is extended to the environment encompassing other organisations, human beings, and most importantly, the natural environment.

Tully (2004) posits that motivation of business, entering an environmental partnership, could be to improve regulatory compliance in a cost-effective way. In other words, environmental partnerships permit companies to get access to information that will allow them to "further refine their environmental risk management systems and reduce environmental liability under national law" (ibid:7). With regard to the NPO's motivation for entering a partnership with a business, they can increase "the impact of their regulatory influence over government by harnessing the commercial influence" (ibid) within the public sector. Hence, their lobbying to the government improves as it becomes more informed and effective through their access to their business partner (ibid).

Oliver (1990) suggests asymmetry as a motive of organisations in the case for collaborations due to their desire to exercise control over the resources of another organisation. In fact, asymmetry refers to the lack of balanced distribution of power between the two parties and indicates power imbalances or the domination of one partner over the other. According to Pfeffer and Salancik (1978) the reason organisations interact with each other is to enhance their control over resources that other organisations control.

Reciprocity is another motive: "motives of reciprocity emphasize co-operation, collaboration, and coordination among organisations, rather than domination, power and control" (Oliver 1990:244). Hence, as it is primarily suggested by the partnership literature (Stafford and Hartman 2001) organisations develop partnerships for the purpose of pursuing mutually beneficial aims. Exchange theory (Hall et al. 1977;

Levine and White 1961) suggests that organisations meet their objectives and aims through the exchanges that they carry out. The need for interorganisational relationship exists because resources are scarce (Levine and White 1961).

In their literature review, Selsky and Parker (2005:7) suggest that "partner motivation is a frequent topic for research in the formation stage because motivational differences are believed to derail collaborative intent". They report that the motives of NPOs tend to be altruistic, unlike the motives of BUSs which predominately pursue self interest (Iyer 2003), including desire to improve public relations, receiving scarce technical assistance (Milne et al. 1996), to enhance the image and reputation of BUSs (Alsop 2004; Heap 1998). Also BUSs can attain credibility (Heap 1998) by associating with NPOs by validating their efforts in addressing social issues as genuine problems. According to Koza and Lewin (2000:256) a company's motivation to enter into an alliance could be to either "exploit an existing capability or to explore for new opportunities". This would usually be centred around the capabilities or expertise of the NPO.

On the other hand the motivation for NPOs to enter into a relationship with a BUS can also vary. For example for a NPO the reasons for a collaborative relationship include: to enhance resources (Wymer and Samu 2003; Milne et al. 1996; Heap 1998; Fishel 1993) and credibility (Heap 1998); to improve access to networks, contacts and technical expertise (Heap 1998) and to facilitate the acquisition of information (Macdonald and Piekkari 2005).

Although the majority of studies assume different priorities among the two partners, Selsky and Parker (2005:8) suggest that more "recent research re-examines that assumption", offering the example of NPOs that attempt to elicit social change through the relationship with BUSs (Fabig and Boele 1999).

According to Covey and Brown (2001:16) there is evidence that partnership relationships are difficult to establish due to the differences between the partners; however, "there is also growing evidence that many stakeholders increasingly see a convergence in rights and expectations for business and civil society" (ibid).

The above observation of a gradual convergence in rights and expectations reflected a much earlier reality which is well-established today both among BUSs and NPOs. The earlier divergence and today's convergence between the sectors are two contrasting trends highlighting the need for more empirical research within the new era of partnerships (Austin 2000; Googins and Rochlin 2000) as accepted forms of cross-sector collaboration and for more examination of the extent to which partners from different economic sectors have distinctive and conflicting agendas.

The book closes this gap by questioning the assumption that the priorities, hence the motives, of the partners are different or conflicting by asking if the motives are shared among the partners. The following research questions are addressed by this research:

RQ3: What are the motives of the NPO and BUS partners?

RQ4: Are the motives between the partners shared?

The next section discusses within the literature the process of partnership building and the dynamics of partners within the implementation phase of partnerships.

2.4.2   Partnership Implementation

The partnership implementation refers to the interactions of the partners within the partnership relationship. The literature suggests that partnerships are resource intensive relationships with increased interactions among the partners (Austin 2000). Furthermore there are differences in the structure these relationships can take ranging from formal agreements (Austin 2000) to informal loose collaborations (Berger et al. 2004).

Googins and Rochlin (2000:133) refer to the 'critical steps' within the process of partnership building and suggest six steps: (1) defining clear goals; (2) obtaining senior-level commitment; (3) engaging in frequent communication; (4) assigning professionals to lead the work; (5) sharing the commitment of resources; and (6) evaluating progress/results. On the other hand, Andrioff (2001:224) refers to the four Ps of stakeholder partnership building: the purpose of partnerships, the pact between the partners, the power relationships within the partners and the process of the evolution of partnerships. In his study of four stakeholder partnerships he identified different characteristics for each partnership relationship under the four Ps. A number of either prescriptive or descriptive steps of partnership-building exist within the literature (Berger et al. 2004; Wilson and Charlton 1997; Westley and Vredenburg 1997) that their common characteristic is their chronological sequence of relationship evolution (Selsky and Parker 2005).

Identifying the process or the phases within the process of partnership-building will permit the observation of the level of institutionalisation of the partnership in each partner organisation, based on the perceptions of the interviewees. The phases of the process are also used as a way to examine the dynamics across the two organisations during the course of the partnership. The respective research question that follows from the above is:

RQ5: What are the phases of the partnership process in the cases under examination?

Partnerships are "diverse in purpose, size and scope" (Googins and Rochlin 2000:134), an observation that has been tested and verified through many case studies (Berger et al. 2004; Andriof 2001; Austin 2000). Similarly each phase of the relationship requires interactions between the partners that permit observation of the dynamics. As pointed out by Googins and Rochlin (2000), as the partnership progresses the levels of dependency increase. Balancing power asymmetries, Hamman and Acutt (2003) suggest, is one of the most significant concerns with regard to the interaction among NPOs and BUSs and frequently the imbalance of power favours BUSs. Fisher and Ury (1981:106) suggest that by developing the "Best Alternative to a Negotiated Agreement (BATNA)" NPOs will be able to decide when it is advantageous to partner with BUS and when it is not. This is in agreement with Covey and Brown's critical cooperation (2001) suggestion, which incorporates according to Hamann and Acutt (2003:263) "rights-based and power-based approaches to negotiations". Hence Uri (1991)) as well Hamman and Acutt (2003) suggest that a/ developing a number of strategies, such as "taking legal action, organising local level protest and solidarity strikes" (ibid:264) and b/ communicating a "warning rather than a threat" (ibid) should entail an "advance notice of danger" (ibid) to the BUS. However, the above can be true in the case of NPOs able to formulate these strategies and more importantly when the BUS partner perceives the NPO as capable of pursuing them. Hence, the previous experience of the NPO and its mission are important contributing factors. Furthermore, large power imbalances (Lister 2000) among the partners that favour the party that dominates financial resources result in a "hidden powerful effect" (Shaw 1993, cited in Millar et al. 2004:406) that hijacks the intention of objectivity or criticism of the less powerful actors. As suggested by Shaw (ibid) "it takes a lot to bite the hand that feeds you: a muzzle is a good insurance against unwelcome bites". On the other hand the decision-making process and structures of NPOs are not usually transparent (Millar et al. 2004) which further colours the internal partnership dynamics as opaque and removed from being integrated within a broader social context.

A significant aspect of partnerships is their closed character as suggested by Rowe and Devanney (2003:378): "Post-Enron, the accounting profession's close relationship with clients has been exposed. In seeking to engage different sectors and individuals in partnerships, to ignore the potentially closed and exclusive character of networks is foolhardy". As non-regulated forms of association NPO-BUS partnerships follow the BUS-BUS partnerships model that is characterised by confidentiality that results in competitive advantage. Rondinelli and London (2003):72) pose the question: "Can both corporation and NPO maintain confidentiality?" They suggest that this is linked to the issue of trust between the two partners, which is extensively discussed in the NPO-BUS partnership literature (Parker and Selsky 2004; Hardy et al. 1998; Iyer 2003; Huxham and Vangen 1996). However, the outcomes of social partnerships, if they are socially embedded and aiming at societal outcomes then wider participation, would be improved by plurality of opinions and open dialogue on the aspects of partnership, and this would also increase the social legitimacy and accountability of the relationship. Zeng and Chen (2003:588) point out that "while trust is essential in promoting cooperation in alliances, an over-trusting partner can become an easy target for exploitation by its greedy partners".

An important aspect that appears to be central within partnerships is the development of a collective identity among the partners. Jenkins (1996) describes "collective identity as constituted by a dialectic interplay of processes of internal and external definition." The sense of belonging and what is usually referred to as 'we' before the formation of the partnership represents the original organisational identity, also recognised by outsiders to the organisation. However, during the partnership formation the reference to 'we' needs to change if indeed a collective identity is to emerge (Lamont and Molnar 2002). The research aims to study the extent to which a collective identity emerges during the implementation of the partnership. Since the two organisations do not merge during the partnership the collective identity is more likely to represent the symbolic 'oneness' that organisational actors hold within their own perceptions based on the processes that take place during the partnership formation. As Lamont and Molnar (2002:182) remark: "Individuals within such categorical communities have at their disposal common categorisation systems to differentiate between insiders and outsiders and common vocabularies and symbols through which they create a shared identity". It is often the case that people who share categorisation systems are considered members of symbolic communities irrespective of the similarities or differences in their way of living (Lamont 1992; Wuthnow 1989; Hunter 1974).

Consequently, the existence of different or similar beliefs, perceptions and assumptions among the partners is important. The literature on NPO-BUS partnerships as presented above suggests an a priori conflict between the profit and nonprofit sectors. If indeed this conflict exists within the partners it should be demonstrated through the partners' dynamics. According to Hatch (1997), the early literature on co-operation views conflict by large as 'dysfunctional' "as it was believed to be the antithesis of cooperation" (Hatch 1997:302). Hence structural mechanisms were suggested as a way to reduce or manage conflict (ibid), such as forms of task forces and committees as forms of coordination (Galbreath 1977). Hatch (1997) remarks that Pondy's work (1967, 1969) moved the examination of conflict into its second phase by accepting it as unavoidable, hence representing a natural condition. This led theorists into searching for explanations of the conditions under which conflict emerges. A third view of conflict as a functional aspect of organisational life presents a challenge for the previous assumptions as Hatch (1997:304) remarks:

> The functional view of conflict proposes that conflict is good for the organisation because it leads to stimulation, adaptation and innovation and better decision making, largely as a result of the input of divergent opinions. ...It is sociologically healthy because it encourages opposition to the status quo and initiates conditions of social change. In addition, some theorists credit conflict with providing the conditions for democracy by acknowledging pluralism and encouraging a respect for diversity. The functional perspective also warns that too little conflict can have negative consequences such as group think, poor decision making, apathy and stagnation.

Furthermore, Tyson and Jackson (1992:51) suggested that "a degree of conflict is necessary for a group to perform at an adequate level". However, they remark that if within a group the norm is that of cooperativeness then it will prevail over other norms such as competition or conflict, for example (ibid:205). Hence if a certain organisation does not accept plurality of opinions or its leadership is not used to manage conflict either internally or externally then it is unlikely that when conflict appears it will be able to manage it in a positive manner. However, Robbins (2005:230) considers conflict as necessary for a group to perform effectively and suggests that if the level of conflict is too low the group is similarly dysfunctional as when it is too high:

> When conflict is at an optimal level, complacency and apathy should be minimised, motivation should be enhanced through the creation of a challenging and questioning environment with a vitality that makes work interesting (Robbins 2005:424).

Although conflict is ordinarily seen as the result of 'troublemakers', as pointed out by Senior and Fleming (2006:224) "in democratic society, this strategy will either cause further, more extreme conflict behaviour or drive the expression of conflict underground". This further highlights that organisations cannot represent or be managed as a unitary whole and hence the plurality of opinions is an inevitable fact of organisational life. Tosi et al. (1994:436) define conflict as: "a disagreement, the presence of tension, or some other difficulty between two or more parties. ... Conflict is often related to interference or opposition between the parties involved. The parties in conflict usually see each other as frustrating, or about to frustrate, their needs and goals". Senior and Fleming (2006:220) examined a number of definitions and suggested four aspects of conflict: (1) it must be perceived by the parties to it, otherwise it does not exist; (2) one party to the conflict must be perceived as about to do, or actually be doing, something that the other party (or parties) do not want – in other words there must be opposition; (3) some kind of interaction must take place; (4) in addition almost all accounts of conflict agree that it can take place at a number of levels: between individuals, between groups or between organisations.

The majority of the NPO-BUS partnerships literature has examined partnerships by viewing conflict as dysfunctional or at best as a fact of life, hence it suggested ways of resolving the conflict. This study takes a contrasting position by suggesting that: (a) by refusing to accept the role of conflict as functional in NPO-BUS partnerships we might be missing very important aspects of these relationships and may be ignoring their role in producing beneficial outcomes for society; (b) the plurality of organisations, their characteristics in the organisational context has moved rapidly to new behaviours that reflect the interpretations of their institutional environment. Hence, although in the past boundaries between the sectors, and the clear differentiation of the roles and responsibilities of organisations suggested finding ways to collaborate in order to increase the understanding and collaboration, today the increased interactions result in pronounced isomorphism (Kolk et al. 2008; Heap 1998; Galaskiewicz and Wasserman 1989) across different sectors and organisations might call for research to address the question of whether there is difference and functional/productive conflict within partnerships.

If indeed differences exist between the two organisations then conflict should occur within the relationship manifested in local conditions or in other words within the relationship dynamics. Based on Hatch's interunit conflict model (1997:308–313) examples of local conditions of conflict that can be applied beyond the interunit model can include: (1) group characteristics based on the differences that exist across different organisations or units within the same organisation; (2) goal incompatibility that stems from the different mandates and goals of organisations; (3) task interdependence – Hatch suggests that the increased interaction across different units in order to perform tasks together can present multiple opportunities for conflict or indeed the mutuality on task dependence of the partners might moderate or hinder the conflict; (4) common resources – the dependence on a shared or common pool of resources can often provoke conflict; (5) status incongruity – the imbalance of status within or across organisations can cause conflict in cases of inversion of status hierarchy (i.e. if a lower status group influences the activities of a higher status group); (6) jurisdictional ambiguity – when responsibility either for credit or blame is not clearly delineated across the groups this might create opportunities for conflict; (7) communication obstacles – when different groups "speak different languages" (Hatch 1997:312) they find it difficult to agree on issues of mutual interest; (8) individual differences that stem from personality conflict; and finally (9) rewards and performance criteria which offer opportunities for comparisons across groups with regard to the performance of organisations or groups can create conflict among the groups.

As suggested earlier, the literature of NPO-BUS partnerships by and large fosters a strategic perspective in the examination of cross-sector relationships, hence conflict is viewed as dysfunctional. This study views conflict as part of the process of social interaction where "ideas, beliefs and pre-assumptions are challenged vigorously" (Bennett and Savani 2004:182) contributing to positive outcomes for the partners and more importantly for society. Hence, the aim here is to examine if conflict exists between the partners during the implementation phase and the dynamics of the partnership relationship. The research questions addressed are:

RQ6: How do the dynamics between the two partners evolve through their interactions?

RQ7: Is conflict observable within the implementation and through the power dynamics of the partners?

The next section discusses the outcomes of partnerships for both the respective partners and for society at large.

2.4.3 Partnership Outcomes

Heap reported as early as 1998 that the impact of the private sector on NPOs resulted in a number of changes: changes in the relationship between the two sectors, changes in the delivery of programmes, changes in the organisational systems and structures and changes in the organisational forms. As Heap (1998:8–9) remarks:

> While statutory legal and tax systems keep NGOs and the private sector apart, the cross-fertilisation of each other's vocabulary and methods is making sectoral frontiers increasingly blurred, a breeding ground for hybrid for profit/nonprofit organisations (Leat 1993; Davis 1997). INTRAC's own millennium paper lays out this institutional isomorphism as 'NGO Incorporated': …we have identified…the dramatic change in the nature and form of NGOs. Even the language used to describe NGOs is changing and identities and boundaries are clearly shifting. We see NGOs with no value base, as against commercial enterprises with very strong ethical values; certain nonprofit organisations run counter to the traditional spirit of voluntarism; commercial consultancy firms competing with established NGOs to operate programmes in former 'NGO territory' and NGOs setting up commercial consultancy wings (INTRAC 1997).

The organisational forms are still captured in the traditional separations of profit and nonprofit although in reality the ways in which they operate have shifted (Heap 1998). Millar et al. (2004:410) also emphasise the duality in the identities of NGOs: "a market as well as an institutional identity in today's global business environment". Within these institutional changes it is important not only to look at the outcomes NPO-BUS partnerships can deliver for each participating organisation but also for society in order to understand better the impact of the institutional changes.

The literature on NPO-BUS partnership makes an implicit distinction between organisational and social outcomes. For example, Kanter (1999) makes reference to 'tangible business benefits' and the need for 'new knowledge and capabilities that will stem from innovation', one of the potential positive outcomes from the partnership relationship and presents examples of how these relationships even turned out to be financially beneficial to BUSs. Drucker (1989) suggests that an important and unexpected positive outcome for BUS is the new management practices BUS can learn and adapt from NPOs which have experience with multiple bottom lines, similar to the recent triple bottom line perspective that BUS need to attend to. Furthermore, energy reduction, environmentally friendly production and service provision have been marked as positive outcomes (Rondinelli and London 2003). These present borderline organisational/societal outcomes as they benefit the organisation but also society due to the reduction in environmental disturbance.

As it becomes obvious from the above, there is a category of outcomes that can be classified as intangibles, including knowledge and capabilities. The intangible assets were considered as early as 1987 to be the most important resource for a company (Itami and Roehl 1987), one expression of which is the core competencies (Prahalad and Hamel 1990). Organisations, predominately BUSs, actively engage in acquiring or internally producing intangible resources as they are likely to increase the value of the company (Sanchez et al. 2000). As Galbreath (2002:116) points out, one of the most far-reaching changes in this field in the twenty-first century concerns what constitutes value and what the rules of value creation might be. Moving from the tradition of tangible to intangibles and to relationship assets constitutes a change in perceiving where the value of the firm is positioned today: "what becomes easily apparent is that the firm's success is ultimately derived from relationships, both internal and external" (Galbreath 2002:118). Furthermore, Galbreath reports that more than 20,000 alliance partnerships were formed in BUS within 2 years (2000–2002) worldwide and that a "typical large company manages 30 or more alliances" (ibid:122) which testifies to the extent of experience BUSs have in partnering. The relationship between NPOs and BUSs is seen today as a source of cross-sector intangible outcomes that can benefit all parties (Seitanidi 2007b). It is important, however, to distinguish between the different types of outcomes with regard to who is the recipient of the benefits.

Caplan (2003:34) makes an important distinction between outputs and outcomes of partnerships. He suggests that through a partnership more water points can be installed in a community. However partnerships do not provide a unique mechanism through which water points can be installed. Alternatives include donors providing the funding for either a BUS or an NPO to install water points or alternatively a company offering support to an NPO through any other form of community involvement in order to increase the number of water points. He posits the differences between the two as follows:

Outputs are tangibles that we can see – a report, the number of times an advocacy message is repeated on television or radio, the number of children vaccinated, etc. We tend to stop after we have counted all the outputs of the partnership (which in fact may have been completed as suggested above by other means even quicker) (Caplan 2003:34).

Outcomes are less tangible results – how many children can now attend school because they are not walking 4 miles a day for water; how policies have been changed in a company or legislation amended as a result of findings documented in a report; how behaviour has changed as a result of hearing a message repeatedly; or estimates of how much money has been saved in curative care as a result of vaccinations being given to children (ibid).

The presumption of social partnerships is that by combining the resources, expertise and synergies of the two partners the partnership is able to contribute to society in a unique way. Following from this, another distinction needs to be made: there is a difference between the intentional or unintentional outcomes, highlighting the importance of intent. A number of authors suggest that setting goals from the outset (Austin 2000; Andrioff 2001; Wilson and Charlton 1997) and assessing a number of measures of fit between the two organisations is a way of achieving better outcomes (Berger et al. 2004). However, the partnership cases presented in the literature indicate that the added value in partnerships derives from the process itself and the unintentional outcomes for the organisations (Seitanidi and Ryan 2007; Austin 2000). The above is possible as the two partners are present within the relationship; society and the environment on the other hand as 'silent partners' are represented through the NPOs. Within the changing institutional environment of increased sectoral isomorphism it might prove difficult to deliver outcomes that are beneficial for society.

Caplan and Jones (2002:1) remark that in order for the partners to continue to be interested in the partnership the relationship has to respond to their needs and deliver value to the organisations and the indicators of the partnership, or in other words the partnership outcomes. However, if indeed indicators are set up to measure the individual and mutually beneficial outcomes similarly the beneficial outcomes for society should be taken into consideration in the case of social partnerships. In certain instances the organisational goals (hence outcomes) might be in conflict with the benefits for society. In this case the absence of any third-party involvement, such as the government or another NPO or a group of NPOs, might jeopardise the credibility of the relationship with regard to serving the public good. The failure of partnership to focus "openly and clearly" (Caplan 2003:34) in the societal outcomes can contribute not only to the disappointment of the partners (ibid) but more importantly to decreasing public trust in the ability of the partnership institution to deliver the promised 'social goods'. The measurement of the effectiveness of the relationship as suggested by Caplan and Jones (2002:2) plays an important role as it allows partners to determine "what works less well and make adjustments. They will be able to minimise the costs and maximise the benefit of the partnership". Consequently the question to be asked is 'maximising the benefits for whom?'. Hence it is important to look at both the organisational and societal outcomes and not just at the social outcomes that derive from the work of the NPO.

Within the book when reference is made to 'outcomes', organisational outcomes are implied, directly linked to the benefits that derive through the partnership for the organisations. In the case of NPOs, since their missions serve social ends, the outcomes that occur are referred to as 'social outcomes' (what Caplan terms 'outputs'). On the other hand 'societal outcomes' refer to the unique benefits that accrue for society through the partnership relationship. It is believed that the above terms indicate more clearly the distinctions rather than the suggested 'outputs' and 'outcomes', terms used by Caplan (2003) for the purposes of this study. The distinction suggested here points to the difference of outcomes that derive from the NPO only versus the outcomes that derive through the partnership relationship.

The organisational outcomes for the profit sector can include reputation enhancement; improving their image and credibility through the association with NPOs which can in effect result in positive financial benefits (personnel retention and attraction, risk aversion for boycotts; and acquiring intelligence based on NPOs areas of expertise (Heap 1998:17–22). Greenall and Rovere (1999:3–5), in their research within the Canadian context, suggested that the engagement of BUS in partnerships is a necessity that results in: securing access to land; trust and reputation building and assurance; social licence to operate; access to local community users of corporate goods and services; marketing and competitive advantage; improved quality of decision making; and overall industry health. Another important outcome for BUS that they reported is that partnerships "allow quicker, more direct and comprehensive access to the feelings of stakeholders (and, therefore, more rapid and solid stakeholder consensus" (ibid:8).

The organisational outcomes for NPOs include access to networks and contacts that they previously did not have access to, technical expertise, management expertise, financial discipline and customer orientation in the delivery of their services (Heap 1998:23–26). According to Austin (2000:88) some of the positive outcomes for the NPOs include: "financial resources, services or goods, technical expertise and technologies, access to other corporations, enhanced name recognition, and new perspectives". Following from this it is important to differentiate between the organisational and societal outcomes in order to qualify the relationships as social partnerships by following Waddock's definition (1988). The research question that the research addresses is:

RQ7: What are the positive outcomes that accrue from the partnership relationship in the cases under study for the organisations and society?

Within the literature regarding the partnership outcomes, change appears (implicitly and explicitly) to be both an intentional and an unintentional (hence processual) outcome for both partners on the organisational level but also for the sectors they represent on the institutional level. Reference has been made to structural changes that take place within both sectors as a result of the intense interactions (Millar et al. 2004; Heap 1998) but also on the participating organisations (Loza 2004; Ebrahim 2005). Nevertheless, predominately the literature does not address the issue of intention within NPO-BUS partnerships (Seitanidi 2008). Although within movement and mobilisation theory and in the case of confrontational cross-sector relationships attention has been given to the intention for change, in collaborative relationships the central issue of intent has been largely ignored apart from exceptions (see Bendell 2000c). NPO-BUS partnership case studies discuss the changes that take place, predominantly on the organisational level, however it is not clear who initiated the changes and if and how they were followed through strategically within the partnership relationship. Examining the intention within partnerships and in particular the intention for change offers an indication of the ability of the organisations from different sectors to successfully follow through their intentions (originally expressed as motives), to acquire the means of achieving them (implementation process) and to arrive at beneficial outcomes.

Since change is a broad research area where many different theoretical approaches have been employed, offering a literature review on change would be beyond the aims of the chapter. The purpose here is to point out the shift that took place over recent years in the study of change, but also to highlight theoretical issues that touch upon the phenomenon under study i.e. NPO-BUS partnerships and its relationship to change.

Perrow (1994) has argued that "explaining change is – or should be – a central concern of organisational scholars today" (cited in Sastry 1997:237). As Shen points out "organisations are agents of change" (Shen 2005:3) facing pressures from both their internal and external environment in order to increase their efficiency, effectiveness "usually leads to planned, organisation-wide change, such as transforming 'market empires' into 'relationship masters' (*e.g. Enron*[4]). Organisational transformation is a mode of social change that involves a sharp and simultaneous shift in strategy, structure, process and distribution of organisational power" (Shen 2005:3). In fact, according to Martin (2000:452), "to change is to take different actions than previously. To take different actions than previously means to make different choices. Different choices produce change. The same choices produce sameness, a reinforcement of the status quo". As he continues he remarks that "to espouse a different operating principle (e.g., we have decided to become customer focused) from the past does not represent change. Only if different choices lead to action on the different operating principle will change be produced". However, since "…organisations have developed protective devices to maintain stability and that they are notoriously difficult to change or reform should not be allowed to obscure their dynamic relationships with the social and natural environment. Changes to the environment lead to demands for change in the organisation, and even the effort to resist those demands results in internal change" (Katz and Khan 1978:30–31).

Pettigrew (2000:246) remarks: "change and organisation are static nouns, whilst Weick argues that we need a dynamic vocabulary of changing and organising if we are to take charge of a changing world". In 1985, Pettigrew critiqued the research on organisational change and described it as "largely acontextual, ahistorical, and aprocessual" (Pettigrew 2000:243). In a later paper, he commented "research and writing on organisational change is undergoing a metamorphosis" (Pettigrew et al. 2001:697).

---

[4] Insertion in italic mine.

As he remarked in the recent years more scholars are concentrating on the aspects of continuity in change (Greenwood and Hinnings 1996; Van de Ven et al. 1989) acknowledging that "context and action are inseparable" (Pettigrew et al. 2001:697). Looking at the work of Pettigrew (2000) we can chart the literature on change. Greenwood and Hinings focus in their paper on the issue of the interaction of organisational context and organisational action. Although institutional theory and the punctuated equilibrium were two views that influenced research on change (Pettigrew 2000), they recently attracted scholarly criticism (Sastry 1997; Greenwood and Hinnings 1996). Furthermore, according to Pettigrew (2000:243) another interesting addition is the research of Brown and Eisenhardt (1997, 1998) who "seek to blend evolutionary theory and complexity theory in order to expose and explain how innovation may occur 'at the edge of chaos'". Lewin et al. (1999:535) argue that "firm strategic and organisation adaptations coevolve with changes in the environment (competitive dynamics, technological and institutional) and organisation population and forms, and that new organisational forms can mutate and emerge from the existing population of organisations".

As mentioned in the above studies (but also in the research of Dutton et al. 2001; Heracleous and Barrett (2001); Lovelace et al. (2001)) change is conceptualised as an interaction field: "focusing on interaction moves away from the variables paradigm toward a form of holistic explanation. The intellectual task is to examine how and why constellations of forces shape the character of change processes rather than "fixed entities with variable qualities" (Abbott 1992:1). Rather than causality being attributed to variables, social actors move onto the stage of history as agents of history. Change explanations are no longer pared down to relationships between independent and dependent variables but instead viewed as an interaction between context and action. Crucially, context is used analytically not just as a stimulus environment, but also as a nested arrangement of structures and processes in which the subjective interpretations of actors' perceiving, learning, and remembering help shape process" (Pettigrew et al. 2001:699).

In his comprehensive review of the NGO literature, MacKeith (1993) points out that "only 11 articles refer directly to the issue related to organisational change and growth of NGOs" (Dechalert 2002:15). As remarked by Dechalert, although the NGO literature has increased and issues around organisational change have been raised, since 1993, it has not been investigated in depth apart from the recent works of Ebrahim 2005 and Lewis 2000. On the other hand, change has been researched extensively within the business management literature (Poole et al. 2000; Pettigrew 1990; Argyris 1985). Finally, the NPO-BUS partnership literature has not previously looked at partnerships as agents for change but rather focused on NPOs facilitating or forcing change within BUS (Bendell 2000c). Hence this research attempts to close this gap in the literature and offer an insight into how partnership among the two organisations can deliver change either as an intentional or an unintentional outcome for either partner but also as part of the process. The final research question that the book addresses is:

RQ8: Is change an intentional outcome for either the profit or the nonprofit partner as a result of the partnership relationship?

2.5 Conclusion

The second chapter of the book reviewed the literature on NPO-BUS partnerships in order to position the study and its contributions. It identified six strands within the literature (nature of partnership, managerial aspects of partnership, strategic use of partnership, legal and ethical consideration in partnerships, partnership measurements, societal implications) and three perspectives (single, dual and societal) and two types of analysis (critical and uncritical). The current research is positioned within the sixth and most recent strand of the literature studying the societal implications of NPO-BUS partnerships by studying the nature and managerial aspects on the organisational level and suggesting inferences on the societal level. The study fosters a societal perspective by attempting to include in the analysis not only both partners but society by discussing the implications of the relationship on the societal level. It finally adopts a critical perspective towards the phenomenon by questioning a number of assumptions that were previously taken for granted within the literature.

The three stages of partnerships, namely: formation, implementation and outcomes have been chosen to group the theoretical issues within partnerships. Under the first stage the role of the organisational characteristics of each partner, the importance of the historical evolution of the relationship between the partners and the motives of each partner organisation were discussed within the literature. Under the second stage of partnership implementation both the process and the main issues surrounding the relationship dynamics were considered. The last section suggested a distinction on outcomes: organisational, social (those that accrue from the NPO), and societal. The latter are the unique combined outcomes of the partnership's efforts and they determine if the form of association is a social partnership. Each section presented the research questions that the book examines within the remaining chapters of the book. A holistic framework that encompasses all three phases was put forward in order to study the phenomenon of partnerships. The chapter suggested that in today's organisational reality the previous differences between organisations from different economic sectors appear to diminish and hence it might be the case that divergence has replaced by convergence between the different sectors and respective organisations resulting in sectorial isomorphism. Following from the above the differences that led to functional conflict might not be present in certain cases. Hence by examining in three stages NPO-BUS partnerships the research aims to shed new light not only on the partnership phenomenon but also on its implications for the participating sectors.

Chapter 3
Stage One: Partnership Formation

3.1 Introduction

The chapter presents Earthwatch and Rio Tinto, The Prince's Trust and the Royal Bank of Scotland within the partnership case studies. Each relationship is introduced through a brief overview of the partnership. The organisational characteristics of each case study organisation are highlighted in order to identify the types of NPOs and BUS participating within each partnership relationship. Following from the above the evolution of the relationship is presented through the historical perspective of the interactions. Furthermore the motives of each partner organisation are analysed and compared within each case study in order to address whether the motives are shared between the partners. The discussion of the chapter compares the findings of the two in-depth case studies with 35 interviews that were conducted within 29 organisations in order to confirm or disconfirm the findings on the formation stage of partnerships.

3.2 Partnership Overview

As an introduction to the case studies a brief overview of the partnership between the organisations is presented in order to introduce some aspects of the relationship. A more detailed description of the Earthwatch (Earthwatch) and Rio Tinto (Rio Tinto) partnership content is presented in Appendix 3. Similarly, a detailed description of the partnership between The Prince's Trust (PT) and the Royal Bank of Scotland (RBSG) is presented in Appendix 4.

M.M. Seitanidi, *The Politics of Partnerships: A Critical Examination*
of Nonprofit-Business Partnerships, DOI 10.1007/978-90-481-8547-4_3,
© Springer Science+Business Media B.V. 2010

3.2.1   Earthwatch-Rio Tinto Partnership Overview

The relationship between the two organisations started in 1990. In 1995 a series of events led to a shift in the associational form of the relationship from a transactional (i.e. sponsorship) to an integrative (i.e. partnership) relationship. The year 1995 is identified as a significant period for both organisations; the chapter sections that follow make reference to the events that contributed, in my opinion, to this shift in the associational form based on the material at hand.

The negotiations for establishing a partnership started in 1997 and 2 years later the two organisations signed the first Memorandum of Understanding (MoU) between them which signposted the beginning of their first global partnership in 1999. The first partnership lasted 3 years and in 2002 they signed the second global partnership, which lasted until 2004. Five years later the partnership between Earthwatch and Rio Tinto continues testifying the strong relationship between the organisations that has lasted so far for 19 years.

The partnership between Rio Tinto UK and Earthwatch Europe consists of seven components: (1) Rio Tinto is a member of the Corporate Environmental Group (CERG) of Earthwatch; (2) the Rio Tinto Employee Fellowship Programme, whereby 24 Rio Tinto employees are selected every year to participate in six Earthwatch biodiversity research projects in countries where the company operates; (3) field project grants are given from Rio Tinto to Earthwatch to support the field projects that its own employees participate; (4) the Annual Lecture Series, organised by Earthwatch and supported by the company; (5) the African Fellowship Programme, supporting African conservation professionals; (6) core funding is provided by the company in order to support Earthwatch's science department – Rio Tinto also covers the costs of a part-time position for the co-ordination of the partnership; (7) Project Development. Rio Tinto covers the costs for the development of new Earthwatch field projects, specifically for the company, in areas of mining or exploration, involving Rio Tinto employees as volunteers.

3.2.2   The Prince's Trust-Royal Bank of Scotland Partnership Overview

The relationship between the two organisations started in the 1980s. In 2000 after the takeover of Natwest and the formation of the Royal Bank of Scotland Group a series of events led to a shift in the associational form of the relationship: from transactional (i.e. sponsorship) to an integrative (i.e. partnership) relationship. The year between 2000 and 2001 is identified as a significant period for both organisations; the sections that follow make reference to the events that in my opinion contributed to a shift in the associational form.

The first partnership lasted 3 years; it started in 2001; the total amount of money that was given to the PT within the course of the partnerships was £3.4 million. In 2004 the second partnership commenced; the total amount of money given to the Trust was £5 million within a time frame of 5 years.

The partnership between PT and the RBSG consists of seven components: (1) the funding of the initiative 'Route 14–24' which was formed as a result of the partnership; (2) RBSG's employee involvement allowing the PT to increase its volunteers resources; (3) the establishment of the 'Business Awards', a recognition event for young entrepreneurs throughout the country; (4) the funding of 'Business Start-ups', by disadvantaged young people; (5) the encouragement of donations by wealthy individuals through the relationship with Coutts Wealth Management division of the RBS Group; (6) sponsorship of events, in particular sport e.g. 'The PT Cricket' initiative; (7) seconded IT resource from RBSG to PT to assist with the development of critical systems.

3.3 Organisational Characteristics

The term 'organisational characteristics' describes the main characteristics of each partner organisation contributing to the evolution of the relationship and ultimately to the formation of the partnership.

Each of the two partner organisation is presented below. The first section exhibits the organisational characteristics of Earthwatch and Rio Tinto during the pre-partnership years (including the sponsorship years: 1991–1995, the partnership formation years: 1995–1998), and also during the time of the partnership (the partnership years: 1999–2004). The second section concentrates on the organisational characteristics of PT and the RBSG during the pre-partnership years: 1980–2000, but also during the time of the partnership: 2001–2003 in order to allow for differences through time to become apparent.

3.3.1 Earthwatch-Rio Tinto Organisational Characteristics

The main characteristics of Earthwatch are presented below followed by the description of Rio Tinto in order to introduce the two organisations.

In 1973 Earthwatch US founded the Earthwatch Institute network of organisations. Today, it is a global nonprofit organisation with independent branches in Australia (established in 1982), in Europe (established in 1985), and in Japan (established in 1991). Earthwatch Europe was registered as a UK charity in 1985 and opened its Oxford office in 1991 (Earthwatch 2002:46).

Earthwatch Institute Europe has 37 staff[1] members (at its Oxford office), a presence in 118 countries around the world,[2] funding a wide range of projects. The organisation's mission to "engage people worldwide in scientific field research and education to promote the understanding and action necessary for a sustainable

---

[1] Figure from Earthwatch's Annual Report of 2001.

[2] This research focuses on Earthwatch Institute Europe and not on the whole network of Earthwatch organisations. Hence, the word "Earthwatch" refers to Earthwatch Institute Europe, unless otherwise stated.

environment" (Earthwatch 2004c) along with the political position of the organisation as a "non-political, non-confrontational, non-campaigning organisation" places it among the obvious partners for almost any business from every sector. In fact 56% of Earthwatch's total revenue in 2003, which was £16,274,241 (Earthwatch 2003:15), is received from the private sector.[3] According to the model proposed by Dreiling and Wolf (2001) for locating Environmental Movement Organisations (EMOs), organisations that rely heavily on external sources of financial support (such as corporations or foundations) and which are at the same time conservationist[4] are highly collaborative in their approach towards BUSs.

In particular, Earthwatch[5] has found a successful niche in its pro-engagement position towards businesses, attracting many corporations[6] who provide the funding for more than half of the organisation's revenue:

> Earthwatch Institute believes that it is only by actively involving the private sector that we can mobilize the resources necessary to conserve our environmental and cultural resources. Partnership with Earthwatch Institute is a public endorsement of the values, which Earthwatch Institute represents: that objective non-confrontational science should be the basis for understanding and managing the environment.Where companies are supporting programs with Earthwatch Institute, we do not simply ask them for donations. Rather, we seek to actively engage those corporations and their people in strategic partnerships that further both our goals, and involve new community relations, human resource, staff education, leadership development, and communication strategies. (Earthwatch 2004a)

Although Earthwatch presents itself as a non-political organisation, the above statement implies a political ideology[7] that shapes the interpretations of the institutional environment (Basler and Carmin 2002) within which Earthwatch operates. In other words, one of Earthwatch's beliefs appears to be that the conservation of natural resources is going to take place through the mobilisation of the private sector's resources. It appears that Earthwatch's 'sensemaking lens' (Carmin and Basler 2002) does not include the type of actions that would challenge the existing institutional corporate order. More specifically, since Earthwatch aims to influence

---

[3] The name of this category in the annual report was: "grants, partnerships and other income". Also organisational actors interviewed mentioned that over 50% of Earthwatch's income derives from the profit sector.

[4] Conservation groups, such as Earthwatch, "advocate natural resource conservation and wise use of resources and work with a much less inclusive social and ecological identity" (Dreiling and Wolf 2001:41).

[5] Earthwatch as a conservation grant-making organisation is well known for its programmes, which involve volunteers from "diverse countries, cultures and organisations, inspiring them to take responsibility for the environment". Since the focus of this research is the phenomenon of partnerships it is not possible to provide detailed information on Earthwatch's programmes. More information can be obtained from: http://www.earthwatch.org/europe

[6] Earthwatch's Corporate Environmental Responsibility Group (CERG) has 40 corporate members, each paying a subscription of £5,000 per year. Members include: BP, British American Tobacco, Diageo, GlaxoSmithKline, McDonalds, 3M UK, Novartis, Rio Tinto, Shell and Unilever.

[7] "Ideology is a system of meaning that couples assertions and theories about the nature of social life with values and norms relevant to promoting or resisting change" (Oliver and Johnston 2000:40). Also, political ideology largely concerns itself with how to allocate power and to what ends it should be used.

people and their opinions towards conservation it has an inherent political identity which prioritises the role of businesses, as economic institutions, placing them at the focal point of achieving the organisation's aims.

According to Basler and Carmin (2002) the two aspects that shape organisational identity in the case of environmental movement organisations (EMOs) are: (1) the core values that organisational members have towards the environment, which guide the tactics they employ; and (2) the political ideology, which shapes the way in which the political institutions and the political environment within which are interpreted.

Although today Earthwatch is a highly-acclaimed, award-winning organisation, when the relationship with Rio Tinto started in 1991

"Earthwatch was a small and not very well-known NGO and it wanted to become larger and better-known for several reasons" (Interviewee, Earthwatch).

In fact, according to an Earthwatch interviewee, it was the advice of the organisation's bank to develop a corporate membership group in order to attract financial resources:

"look you're not financially viable, you know. Why don't you go and get some corporate money and a membership scheme is a really good way of doing that" (Interviewee, Earthwatch).

Table 3.1 summarises the organisational characteristics of Earthwatch, described below.

Table 3.1 Organisational characteristics of Earthwatch

Attributes	Values
Founding year	1985
Societal sector	Nonprofit
Industry sector	Environmental (biodiversity-conservation)
Mode of operation	Grant making to researchers, engagement and education of multiple publics and voluntarism promotion
Number of staff (UK)	37
Revenue (2003)	£16, 274, 241
Size	Small, entrepreneurial
Mission	Engage people in field scientific research & education to promote understanding & action towards sustainable environment
Headquarters	London (& Melbourne, Australia for Earthwatch Australia)
Strategy of interaction with BUS	Collaborative
Ideology	Non-confrontational, non-political, non-campaigning organisation
Reputation (at formation)	Medium/neutral
Locus of control	Autonomous
Scope of activities	International

At the time the relationship with Rio Tinto was about to begin, a new low-profile organisation, Earthwatch, with lack of funds, which was non-political, non-campaigning and non-confrontational, was carving a niche in a well-established environmental arena with distinctive environmental NPOs on all fronts such as WWF, Greenpeace and Friends of the Earth.

On the other hand "Rio Tinto[8] was formed through the merger of the economic interests of the former "The RTZ Corporation PLC" and "CRA Limited" in a dual listed companies structure. In December 1995 and in June 1997, both companies' names were changed to Rio Tinto" (Rio Tinto 2002:2). As a result the assets of both companies are now managed on a unified basis even though they continue to be separate legal entities with separate share listings and share registers. According to Rio Tinto, "the principal objective to the merger was: "to create a structure to capitalise on future global opportunities, to maximise competitive advantage and to benefit all shareholders of both companies" (Rio Tinto 2002:3). Rio Tinto's management structure facilitates a clear focus on business performance and the Group's primary objective of creating long-term shareholder value in a responsible manner" (Rio Tinto 2002:3). The headquarters for the Group are in London, and Melbourne, Australia is the representative office providing support for the Australian operations of the company.

Production is based mainly in developed countries, primarily Australia and New Zealand; however, the biggest geographical markets are the United States and Western Europe (Rio Tinto 2002:26–47). Rio Tinto's adjusted earnings for 2003 were $1.4 billion and its operating cash flow $3.5 billion with net earnings of $1.5 billion and a total of 36,141 employees worldwide[9].

According to the FTSE4GOOD inclusion criteria Rio Tinto, as a mining company, belongs to the 'high impact' sectors (FTSE4GOOD 2003:1); equally as far as it concerns the human rights criteria it belongs to the 'global resource sector' and to the 'significant involvement in countries of concern' (ibid) (i.e. where the company operates), which are considered companies with the most sensitive inclusion criteria. Although Rio Tinto did not satisfy the FTSE4GOOD inclusion criteria, however it is included in the Dow Jones Sustainability Index.[10] As a mining company it operates within a highly controversial industry, and has been at the centre of the media's attention for its practices in the past. Nevertheless, Rio Tinto, an old corporate name[11], "…has been (for investors) an efficient manufacturer of metal, a consistent generator of earnings growth and dividends, a stock you could buy for

---

[8]"Rio Tinto is a world leader in finding, developing, extracting and processing mineral resources. Diversified by both product and geography the Group is strongly represented in Australia and North America with major assets in South America, Asia, Europe, and Southern Africa. Principal commodity goods and products are aluminium, copper, diamonds and gold, energy (coal, uranium), industrial minerals (borates, titanium dioxide, feedstock, salt, talc) and iron ore" (Rio Tinto 2002).

[9]Figure refers to employees in 2000.

[10]The main difference in the inclusion criteria between the two is that FTSE4GOOD excludes 'companies involved in the extraction or processing of uranium' and 'owners or operators of nuclear power stations'. The first criterion was the one that did not allow for Rio Tinto's inclusion in the FTSE4GOOD Index.

[11]The first mention of the company's name appears in *The Times* archives in 1811.

your grandchildren" (Mortished 2003:23). However, according to Mortished, in the 1980s Rio Tinto was synonymous for the wider public:

> with the ugly face of multinational capitalism, a blight on the landscape, a destroyer of indigenous communities. But, despite vigorous campaigns alleging its baleful influence Rio has grown to become the tallest tree (or the deepest pit) in the mining sector. (ibid)

On May 10th 1995, at the Annual General Meeting of Rio Tinto "nearly 70 Friends of the Earth activists forced company directors to answer questions on the potentially devastating environmental impacts of the proposed mining project[12]" (FoE 1995:1). According to the same Friends of the Earth Press Release:

> At the AGM, the corporation admitted that 50% of the remaining coastal rainforest, which is of global significance, would be lost resulting in the extinction of a number of endemic species should the mining project go ahead. (FoE 1995:1)

As early as 1995, Friends of the Earth were targeting large corporate investors, such as insurance companies, through mailing, in order to exercise pressure on Rio Tinto to influence its decision to pull out of the Madagascar project. Another group that has been campaigning against Rio Tinto since 1978 has been the Partizans[13], a sole issue NGO. The organisation characterises Rio Tinto as "the world's most powerful mining company" (Partizans 2003). They regard themselves as one of the leading practitioners of corporate campaigning, shareholder's activism and disinvestment strategies (Partizans 2003). Most significantly, the interviewee from Partizans suggested that they pioneered the practise of enabling community and workers' representatives to directly challenge massive corporate bodies on their home ground. Nearly 100 such representatives have attended Rio Tinto's annual general meetings since 1981, facilitated by the Partizans (Partizans 2003). The company during the 1980s was a strong follower of Friedman's position, which can be observed from the following statement. According to the Partizans, Sir Rodrick Carnegie (Chairman) during the 1984 RTZ's shareholders meeting in London said: "The right to land depends on the ability to defend it" (Moody 1991:13). At that time, Rio Tinto was still following "Friedman's doctrine" (Friedman 1970:22) and the old hard corporate line that gave priority only to one bottom line[14]: profits focusing exclusively on the interests of their shareholders.

The pressures at a local level from indigenous populations, the internal pressure from employees at the numerous sites of operation within the company and the campaigns on both national and international level organised by a number of pressure groups prepared the ground for a strategy shift. The company was faced with criticism and its operations were being endangered, which ultimately was going to result in profit loss, a priority for any company, but particularly for a mining multinational giant.

---

[12] The project that is mentioned here is the controversial proposed mineral sands mining project in Madagascar.

[13] Partizans stands for: "People Against RTZ and its Subsidiaries". RTZ was one of the names of the two Rio Tinto companies before their merger. Partizans are a sole issue NGO which means that the sole mission of the NGO is to focus on Rio Tinto, its practices and operations around the world.

[14] As opposed to the triple bottom line: economic, environmental, social (Elkington 1999).

Table 3.2 Organisational characteristics of Rio Tinto

Attributes	Values
Founding year (as Rio Tinto)	1997
Societal sector	Profit
Industry sector	Mining
Mode of operation	BUS2BUS
Number of staff	36, 141
Adjusted earnings (2003)	$1.4 billion
Size	Big, multinational
Mission	Creating long-term shareholder value in a responsible manner
Headquarters	London & Melbourne, Australia
Publicly trading shares	London Stock Exchange-FTSE100 & Australia Securities Exchange-S&P/ASX 200
Reputation (at formation stage)	High negative
Scope of activities	International

Table 3.2 summarises the organisational characteristics of Rio Tinto. The industry sector, the fact that is a listed company in stock exchange markets and the international scope of the company's operations elevate reputation as a central organisational characteristic. Rio Tinto's choice for a partner whose ideology and strategy towards BUS was non-threatening, aimed at getting a policy input from an NPO in an area that radical parts of the environmental movement were contributing to the increase of primary risk through secondary risk (Power 2004) i.e. increasing the reputational risk for the company. Despite the differences in the industry and social sectors, size, employee numbers and revenues, it is evident that the two organisations are highly compatible based on their scope of activities, the presence of their headquarters and reputational needs. Their organisational characteristics appear to encourage interactions due to the ideology and tactics of Earthwatch towards BUSs.

3.3.2 The Prince's Trust: Royal Bank of Scotland Organisational Characteristics

The main characteristics of The Prince's Trust are presented below followed by the description of the Royal Bank of Scotland in order to introduce the two organisations.

The PT was founded in 1976 by The Prince of Wales. Having completed his duty in the Royal Navy, His Royal Highness became dedicated to improving the lives of disadvantaged young people in the UK, and he founded the Trust in order to deliver on the above commitment. In fact the Trust's first year was funded with The Prince's severance pay from the Navy of £7,400. Today, the Prince of Wales remarks that it was the fact that young people were not given opportunities quickly enough that he aimed at meeting through the PT (Prince Trust 2004a):

> No one was putting the trust in them they needed. If I was going to do anything, it had to be an operation that was able to take those risks: to trust young people and to experiment. (President, PT)

PT has been through a series of transformations since its conception, originally being one charity, which was later broken to five separate charities until 1999, when the various Trust charities were brought together as the PT, while the Queen granted it a Royal Charter (ibid).

The strong links between the organisation and the government are apparent as: (a) 50% of its income is generated from the public sector; (b) its royal character denotes a strong affiliation with the state and (c) its core programmes allow for an alignment with government priorities, such as tackling social exclusion and regional needs across the country. The Trust went through a devolution process to address the need for stronger alignment with the government's regions:

> …when we first started the devolution process … taking responsibilities out to individual regions we restructured the organisation so that it would align with the nine English Government office regions and the three Celtic countries. (Interviewee, PT)

Since The Prince of Wales began the charity, its activities have helped half a million young people move forward in their lives. The Trust's aim for 2004/5 was to support 40,000 young people across the UK, a target which was not only met but also exceeded.

Today, the Trust is the leading youth charity in Britain, offering a wide range of opportunities to young people including training, personal development, business start-up support, mentoring and advice. Operating in nine regions and three 'countries'[15] across the UK, PT has 700 employees and a volunteer force of nearly 10,000.

The main target group of PT is currently the "hardest to reach" (Prince Trust 2003:2), i.e. disadvantaged young people between the ages of 14 and 30 across the country, who have struggled at school, have been in care, had trouble with the law, or are long-term unemployed. The primary objective of the charity is "getting young people's lives working" (Prince Trust 2003a:3) by fostering an "outside the mainstream" (ibid) attitude in order to provide an alternative to the institutional approach of working with young people (Prince Trust 2004b).

An important asset of the organisation's identity is its 'brand' due to the organisational characteristics mentioned above. In 2003 the PT has been through a brand make-over, based on research previously done through a stakeholder consultation process on the charity's audiences, but also employing an agency. The branding exercise resulted in a clearer definition of the way the charity is working by prioritising a set of values embedded in the new brand:

> Enterprising – We are an innovator, with a commercial and independent approach. We have energy and flair and are prepared to take risks.

---

[15] 'Countries' is the preferred term that PT uses for the UK regions.

Enabling – We help people to help themselves by enabling positive change rather than prescribing solutions.
Committed – We are practical and responsible, delivering relevant and reliable solutions. We see things through. (Prince Trust 2003a:5)

Part of the brand's new strategy is the Trust's "ability to create and sustain partnerships with others" (ibid:4), based on the above principles and values. The 'partnership culture' that today is evident throughout the Trust is due to historic reasons of core programme delivery:

So, always we've delivered our Team Programme[16] through that mechanism of partners. (Interviewee, PT)

However, in other parts of the organisation, *the partnership culture* was not present and its adoption was part of a conscious realisation of the need to change:

... many of whom were very switched onto the idea of working in partnership, and I just come from this kind of culture, it seemed, that I was being sort of ... encouraged not to, and that's why I say I felt very isolated and all of a sudden I thought this is, we are going about things completely the wrong way. (Interviewee, PT)

Today, the Charity is delivering its programmes in partnership with a network of nonprofit organisations all around the UK, which means that its "reach is unrivalled outside the state sector" (Prince Trust 2003a:2). The breadth of the charity's services along with its national coverage and the partnership approach constitute part of its organisational identity.

According to the 2003–2004 PT annual report the charity's total revenue was £53,816,000 (Prince Trust 2004c). A financial diversification of the incoming funds (public sector funding 45%, other income from trading subsidiaries, investment and charitable programme fees 29%, charitable trust and other donations 11% and individual donations 4%) is observed, although there is a clear reliance on the public sector funding. The corporate donations amount to 13% of the total revenue or i.e. £7,159,000, which does not seem to create an imbalance such as in the case of the Earthwatch's corporate funding. However, the contribution of the RBSG is almost 52% of the previous figure, which seems to create a lack of diversification within the category of corporate sector funding and hence a potential dependency on the particular corporation.

Based on the above, it is evident that PT was familiar with both the collaborative and commercial approach that allowed it to enter into partnerships with the for-profit sector. Its royal affiliation as well as its previous relations with funders such as high-net worth individuals made it an ideal partner for any high-profile business. The partnership with the Royal Bank of Scotland was not the first such relationship for the Trust, however it was its first integrative relationship with business. Table 3.3 summarises the organisational characteristics of PT, described above.

---

[16] The Team programme aims to develop confidence, motivation and skills through team work at the community for 16–25 year olds (the majority unemployed) within a 12-week personal development course. The course is delivered through partnerships with colleges and fire service departments.

Table 3.3 Organisational characteristics of Prince's Trust

Attributes	Values
Founding year	1976
Societal sector	Nonprofit
Industry sector	Social You Charity specialising in social exclusion
Mode of operation	Youth services and welfare, education and voluntarism promotion[17]
Number of staff (UK)	700
Revenue (2003)	£53, 816, 000
Size	Big, well established
Mission	"Getting young people's lives working" (Prince Trust 2003a:3) by fostering an "outside the mainstream" (ibid) attitude in order to provide an alternative to the institutional approach of working with young people (Prince Trust 2004b)
Headquarters	London
Strategy of interaction with BUS	Collaborative
Ideology	'Enterprising', 'enabling', 'committed'
Locus of control	Autonomous
Reputation	Positive high
Scope of activities	National

Before the relationship with the Royal Bank of Scotland was about to begin, the high profile charity was going through an internal informal dialogue questioning its own approach to the core programmes. The lack of funding over and above its pressing budget that had to be met did not allow for experimentation and change in order for the charity to develop a new working approach and new programmes.

On the other hand the RBSG was founded in Edinburgh by a Royal Charter, in 1727 and is one of the leading financial services providers in the world, employing a total of 126,341 people. Over the years the bank went through mergers and acquisitions of banks such as Coutts and Co, the National Commercial and of National Commercial Bank of Scotland. In 2000 the Royal Bank of Scotland took over Natwest (formed in 1968 by the two biggest high street banks – National Provincial Bank and the Westminster Bank). By the end of 2002 RBSG was the second largest bank in Europe and the fifth largest in the world with branches in Europe, USA and Asia. The RBSG is the leader in corporate banking in the UK, has the largest UK retail network and is the top bank in the country for private banking (RBSG 2004a).

---

[17] According to the International Classification of the Nonprofit Organisations 'Group 4: Social Services', is the category under which PT can be classified.

In order to assess whether pressure was a reason behind the formation of the partnership this study aimed at identifying the criticism surrounding the bank and its operations before and during the period of the partnership.

The Group's annual operating profit for 2003 was £7.15 billion, a record profit, which was characterised as "completely excessive" by the Independent Banking Advisory Service (Flagan 2004). The announcement of the bank's profits prompted an outcry from politicians, the media and consumer groups. The Consumers' Association remarked: "Big Banks are treating customers with contempt. They make these profits through confusing marketing, exploiting the stranglehold they have on the high street" (ibid). On the other hand RBSG's CEO "said that he was: 'not shy about making profit', a high proportion of which, he said, went in dividends to funds which pay pensions to millions" (ibid). The Chairman of the House of Commons Treasury Select Committee, John McFall, "questioned whether banks and credit card companies were taking their social responsibility seriously" (BBC News 2004). In October 2004, a group of MPs investigated the way the four big banks, among which the RBSG, run their credit card business. The bankers agreed to improve their operations so that people would not be "plunged into unsustainable debt" (BBC News 2004). Also the bank was criticised in 2000 by the "Arch-regulator Cruickshank" (Hart 2000) with regard to the Natwest takeover, "the biggest banking takeover" in the UK (eFinancial careers 2003) which he characterised as an "anti-competitive merger in the industry [that] should be prevented" as it "gives it an incredible concentration of power in certain areas, such as Manchester" (ibid).

In June 2003 the bank signed the Equator Principles (EP), which "commit the signatory banks to follow the environmental and social guidelines of the International Finance Corporation (IFC) of the World Bank Group" (Bank Track 2004). More specifically the EP constitute:

> a declaration by several leading banks that they will not provide loans directly to projects where the borrower will not, or is unable to, comply with these principles of socially responsible investment. (RBSG 2004a)

However, according to a Friends of the Earth Corporate Accountability Campaigner:

> The UK banks signed up the Equator Principles but they seem reluctant to be held accountable for their behaviour and the evidence suggests they are failing to put them into practice. One year on, we have to question whether this voluntary commitment to environmental good practice is worth the paper it is written on. (FoE 2004)

The RBSG is one of the banks that the FoE campaign is targeting, due to the funding provided by the bank for the pipeline in Azerbaijan, Georgia and Turkey, which if built, according to FoE, is likely to: "(a) worsen climate change, (b) increase oil spill and (c) cause social strife for local people" (FoE 2004). The campaign urges citizens, but also the bank's clients, to email RBSG's CEO in order to express their concerns.

The RBSG is included in the FTSE4GOOD Index, the Dow Jones Sustainability Index and has signed the UN's Global Compact, to mention only few of the indices it displays in its 2003 Corporate Responsibility Report.

In summary, the main criticism the bank faces stems from: the amount of profits announced, the violation of the EP principles, according to FoE, and also sporadic shareholders' protests that took place over the bonus plans for top RBSG's directors (Hart 2001).

Although the banking sector has been criticised, it seems that the pressures are relatively recent in comparison with those of the extractive industries. Hence, the general public is not equally exposed to the criticisms of the banking sector as it is for extractive industries. However there has been an escalation of criticism from authorities and the nonprofit sector more recently. As a result, from a charity's point of view, the sector can appear of medium risk (based on the FTSE4GOOD classification).

The importance that the RBSG places on its Corporate Social Responsibility is manifested by the fact that the Group's CEO is the designated Board member for the Group's corporate responsibility policy (RBSG 2004a). According to the RBSG its community programme is one of the largest in Europe. The total amount invested in CSR for 2003 was £40.1 million for 2003 (£33.7 million in 2002) (RBSG 2004a). Based on the "Giving List 2003–4", the cash donation of RBSG is £26[18] million, ranking third in the banking sector after Lloyds TSB, with £33.80 million and Barclays with £28.42 million, first and second respectively (Giving List 2004).

The Group's values that guide its CSR programme are as follows: "We have built our business on the principles of honest, openness and integrity and they are the foundations of our Corporate Responsibility strategy" (RBSG 2004a:49) in order: "To deliver superior sustainable value we run our business with integrity, openness and clearly-defined business principles" (RBSG 2004a:46).

An important aspect of the bank's culture is innovation. According to RBSG archives, the bank carries 300 years of tradition in innovation, having invented the overdraft (1728), the first mobile branch (1946), the house purchase loan scheme (1972), the largest IT integration project undertaken in banking world-wide (2002) to mention only a few of its innovations (RBSG 2004a). Risk minimisation is another element that infiltrates the culture of financial institutions that have a tradition in being risk averse and similarly in retaining the local community's confidence. Hence "bankers were held in high regard in their local communities as men of influence and importance" (ibid). The Royal Bank of Scotland in the 300 years of its history went through serial acquisitions and mergers and as a result through many business and organisational changes in order to adapt to the new circumstances. Hence 'change' and 'partnership' are two concepts that constitute fundamental aspects of its culture and therefore shape the organisation. Table 3.4 summarizes the organizational characteristics of the RBSG.

---

[18] According to the company's Annual Report and Accounts for 2003 "the total amount given for charitable purposes by the company and its subsidiary undertakings during the year ended 31 December 2003 was £14.7 million (2002 – £14.7 million) (RBSG 2004b:117).

Table 3.4 Organisational characteristics of Royal Bank of Scotland

Attributes	Values
Founding year	1727
Societal sector	Profit
Industry sector	Banking
Mode of operation	Commercial & BUS2BUS
Number of staff	126,341
Annual operating profits (2003)	£7.15 billion
Size	Big, multinational
Mission	"To deliver superior sustainable value we run our business with integrity, openness and clearly-defined business principles" (RBSG 2004a:46)
Headquarters	London
Publicly trading shares	London stock exchange-FTSE100 & NYSE
Reputation	Medium/neutral
Scope of activities	International

3.4   The Historical Dimension of the Partnership Formation

The following sections present the relationship evolution for both case studies, firstly Earthwatch-Rio Tinto and secondly one the PT-RBSG.

3.4.1   The Historical Dimension of the Earthwatch-Rio Tinto Partnership

Earthwatch Europe was established in 1985, but it was only 4 years later that its first CEO arrived and set up Earthwatch's "Corporate Environmental Responsibility Group"[19] (CERG) in 1990. Rio Tinto joined CERG in 1991 and consequently "was a founding member" (Earthwatch 2002:30) of the group. According to the

---

[19] According to Earthwatch, being a member of CERG means: "Membership of The Corporate Environmental Responsibility Group (CERG) is a public endorsement of the values that Earthwatch represents; that objective science should be the basis for understanding and managing the environment. It is also an opportunity to engage with the environmental movement, to be involved in debate and to share current best practice with our cross-sectoral body of corporate members. Earthwatch has formed a reputation as an excellent partner for business because of its non-confrontational and professional approach, and because it is active in a wide range of disciplines and overseas markets. Earthwatch currently supports 130 field research projects in 50 countries around the world. CERG has around 40 members, drawn from a wide range of industries. Through the Group, Earthwatch engages these companies on environmental issues, seeking to raise awareness within the company. Companies also have the opportunity to develop further programmes in partnership with Earthwatch".

historical information collected for this research, this is the first instance Rio Tinto and Earthwatch Institute Europe embarked on a relationship, as Fig. 3.1 demonstrates. A year later Earthwatch gained Rio Tinto's support for a specific project in Africa, which involved a donation of £15,000 annually for a 4 year period. In 1995, a series of significant events occurred, which were decisive for the course and more importantly the form of the relationship, as presented in the section of the relationship evolution below. At Earthwatch, a new CEO took over, whose experience was from the corporate sector unlike a previous one whose experience was from the charity sector.[20] An interviewee at Earthwatch (1998–2002) describes him:

> …he was somebody who came to Earthwatch from the private sector, and he started forming much more rigorously this philosophy of, if we have partnerships with companies we have an enormous opportunity to change them. And so, although that philosophy under (name of previous CEO) had been latent it was certainly there but it was latent, (name of new CEO) articulated it much more clearly, I think. (Interviewee, Earthwatch)

Meanwhile, Friends of the Earth's campaign against Rio Tinto was at its peak and according to an Earthwatch employee:

> …there was a sense of, in Rio Tinto, 'should we do something about it, and if we should do something about it, what should we do about it?

Since the main pressure and criticism directed to the company was coming from the radical side of the nonprofit sector, Rio Tinto realised it needed to build its reputation by actively being involved with NGOs.

As mentioned above, in 1995 RTZ and CRM merged to form a global leader in mining, Rio Tinto. It was a period of incubation for a new approach for the global mining company. A change in roles within Rio Tinto made the shift possible. One of the changes was a new employee who came to office having worked for the company for many years:

> I was instrumental in setting up the Corporate Partnership Programme which we, I guess, devised out of the UK Corporate Community Giving Programme. (Interviewee, Rio Tinto)

On the Earthwatch side, equally a new director for development came in place who similarly was an experienced Earthwatch employee, and who later became the organisation's CEO. For both organisations the shift from transactional to partnership relationship commenced in 1995 when they embarked in the pre-formation stage of their partnership, Period I, Fig. 3.1, having left behind "The sponsorship years" of the relationship, which ended in 1997.

In 1997 negotiations between the partners started regarding the partnership (Period II, Fig. 3.1). The actual agreement, the first "Memorandum of Understanding" (MoU), was signed in 1998 and in 1999 negotiations finally came to fruition. Earthwatch and Rio Tinto embarked on a 3-year global partnership in 1999, which ended in 2002. This was the first 3 years of a formal relationship between the two organisations (Period III). The second 3-year global partnership, Period IV, started in 2002 and finished in 2004. The blue box in Fig. 3.1 represents the 'Episode' (symbolised with the letter 'E' in the graph) under study, which in the case of this research is the partnership phenomenon.

---

[20] The first Earthwatch CEO came from the nonprofit sector. He used to be the Head of Oxfam UK.

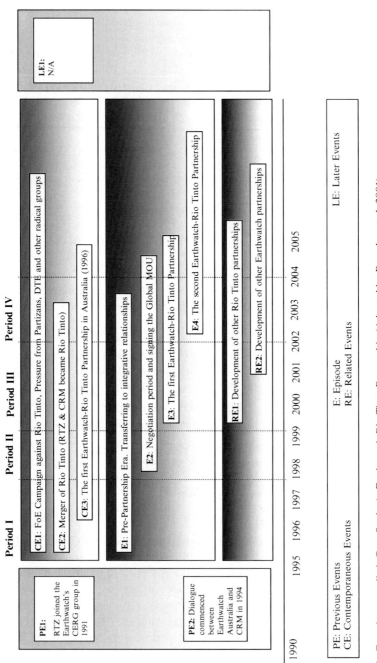

Fig. 3.1 Experience studied: Case Study A: Earthwatch-Rio Tinto Partnership (Adopted by Barzelay et al. 2001)

The green box represents the 'Contemporary Events' (symbolised with the letters 'CE') that happened at the same time as the episode was taking place and the red box the 'Related events' (symbolised with the letters 'RE'). Equally 'Prior Events' are placed in the brown box (symbolised with the letters 'PE' and finally 'Later Events' are in the yellow box (symbolised with the letters 'LE'). Barzelay who developed this model remarks: "Prior and contemporaneous events might be thought of as sources of occurrences, prior events occur before the episode while contemporaneous events occur in the same time" (Barzelay et al. 2001:16). "The related events coincide with the episode and they can be seen as affected by the episode. Equally, the anticipation of later events condition events within the episode" (ibid). The numbers next to the letters, signpost the time sequence of events that take place. As episodes evolve (e.g. E4) chronologically at the same time as the research takes place (e.g. RE1)and since the partnership is still on its course and it is unfolds it impossible to have later events marked as 'LE' in Fig. 3.1. Hence there are no later events and as a result they are marked as 'not available'.

A more concise way, closely identified with the phenomenon of partnership on a macro level and with the historical evolution of the relationship, is to look at the relationship in the following way: (a) The sponsorship years: 1991–1995; (b) The partnership formation years: 1995–1998; and (c) The partnership years: 1999–2004. The first two periods can also be described as the pre-partnership years. Defining three distinctive periods in the relationship continuum is in accordance with the research focus (as the section 'the evolution of the partnership' will examine below), but it will also prove useful for the comparisons that will be drawn between the two case studies.

3.4.2 The Historical Dimension of the Prince's Trust-Royal Bank of Scotland Partnership

The relationship between PT and the Royal Bank of Scotland started before the take-over of Natwest. In fact, the Royal Bank of Scotland, based in Scotland, had a philanthropic relationship with the PT Scotland. Natwest, on the other hand, based in London, had a similar relationship with the PT London office. Hence, for PT's head office (as opposed to the Scottish branch of PT) the relationship with RBSG is relatively new. However, historically the relationship started in the 1980s in Scotland.

> I think … in the early years my experience in The Trust going right back to sort of 1980s, very much philanthropic. (Interviewee, PT)

The relationship between Natwest and PT involved mainly in-kind support through employee volunteering for The Trust.

> … We had a relationship, that relationship was based … was really business mentors. We had a lot of Natwest bank managers or business support managers who supported PT businesses. (Interviewee, PT)

The relationship involved a donation, which was not significant in terms of scale for Natwest and 180 business mentors, volunteering their time for the PT 'business start-up' programmes across the different regions of the UK.

In 2000 the Royal Bank of Scotland had a new CEO who was serving as the Chairman of the PT Board in Scotland. Later the same year the Royal Bank of Scotland took over Natwest bringing the RBS Group to second position in the UK and fifth in the world. At the same time PT was going through structural changes which concluded with the completion of the devolution process in 2003.

According to the Royal Bank of Scotland, after the takeover, RBS commenced a process of integration between the two banks, including the biggest IT integration ever attempted in the world (RBSG 2004a). Part of the integration process was the review of the CSR programmes for both banks.

> ... There was also a link between the RBS in Scotland and because the PT is devolved and there was, if you like, a relationship between London and Natwest and there was a relationship between ... Glasgow and RBS in Scotland. And so, you know, it was effectively joining those four pieces of the jigsaw up. ... and that's when I think the bank was really starting to look at what their corporate social responsibility objectives were, what markets they wanted to operate in, for community investment purposes and then the PT became almost a natural partner I think. (Interviewee, PT)

At the same time the RBSG's CEO hired a new employee who was going to be the "Head of Community Investment", and who in turn set-up a new team of people to work with him. The background of the manager as well as the members of the team was of particular interest. They all constituted *a special breed of executives* who have worked for all three sectors of the economy: profit, nonprofit and the public sector, hence having *multisector backgrounds*.

> But also the fact that (name of employee) and the rest of them in that team they come from either the government or the nonprofit sector maybe that helps the other side to get closer to you. (Interviewee, PT)

The review of the CSR strategy and the addition of the new executives of 'non-traditional banking backgrounds' resulted in a community investment strategy which on the one hand met the vision of the CEO, and on the other was inspired by people who had in-depth knowledge of all sectors but who were also informed of the corporate community investment trends in the USA:

> I mean, yes, you know, I look for people outside, but I was asked to come up with a strategy which was agreeable to (name of CEO) so it's chicken and egg - did I come up with the right strategy that he was happy with, or was it what he wanted... ... Yes, I think that, well one of the differences would be that having worked in the not-for-profit sector myself, I realised that most corporates are very unsophisticated in the way they hand out money. You know, this school of thought, which is primarily driven from the States with venture philanthropy etc. (Interviewee, RBSG)

The decision to develop a structured relationship between the newly-formed group of the Royal Bank of Scotland and the newly devolved PT took place in the higher ranks of both organisations:

> I think probably it was a discussion between (name of director) who's the Director for Scotland, and probably with (name of CEO), or the bank, probably very high level discussion, (name of CEO PT Scotland) knew the way that the Trust would want him to go because of the strategy

and he had a very clear view that that was the right way to go and the bank clearly wanted to do something a bit different. Rather than just give us money for business. (Interviewee, PT)

It was in 2001 that the first partnership relationship was established. The partnership involved the nationwide commitment and support of PT from the RBSG, providing: (1) £3.7 million as a cash donation; and (2) a target of 550 RBSG employees to volunteer in a variety of PTs programmes. At the same time as the partnership decision was taking place, the PT put on tender a loan facility that was going to be used in order to finance the small business start-up programmes. RBSG won the bid for a total loan facility of £7 million, which was a normal business transaction, since there were no preferential terms due to the partnership. However, the press cuttings presented the partnership as the biggest ever in the UK for a total of £11 million.

Meanwhile other partnerships were developed in each organisation either prior to the partnership under examination, simultaneously, or post-RBSG-PT partnership.

Ten months after the interviews were conducted for this case study, the organisations announced a new partnership for the next 5 years (2004–2008). The new partnership, or "new initiative" as it has been termed, involves £5 million with "the aim to help 16–18-year-olds who leave school with no qualifications and 'face poverty of opportunity' in areas of high unemployment" (Prince Trust 2004c).

Although the partnership started in 2001, the previous relationships between PT, Royal Bank of Scotland and Natwest were a contributing factor to the eventual formation of the partnership. Hence, the years from 1980 until the end of 2000 constitute the first period of the relationship, Period I, Fig. 3.2, and "The Pre-partnership years", which include the philanthropic and transactional stage of the relationship. For both organisations, the period from 2000–2001 was an important time as many structural changes took place (Period II). The first partnership commenced in 2001 and ended in 2003 (Period III). This was the first 3 years of a structured relationship between the two organisations. And finally the second partnership started in 2004, after a years discussion among the two partners, and will conclude in 2008 (Period IV).

Figure 3.2 presents the evolution of the relationship across the two partner organisations. The explanation of Barzelay's conceptual framework for the presentation of events was explained in the previous section.

A more concise way, closely identified with the phenomenon of partnership on a macro level and with the historical evolution of the relationship is to look at the relationship in the following way: (a) The pre-partnership years: 1980–2000; (b) The partnership formation years: 2000–2001; (c) The partnership years 2001–2008. Defining three distinct periods in the relationship continuum is in accordance with the research focus, but will also prove useful to the comparisons that will be drawn among the two case studies.

3.5 Partnership Motives

This section concentrates on motives among the partners. In the context of this research, partnership motives refer to the causes or reasons behind the actions of organisations that resulted in the initiation of a relationship action and ultimately to

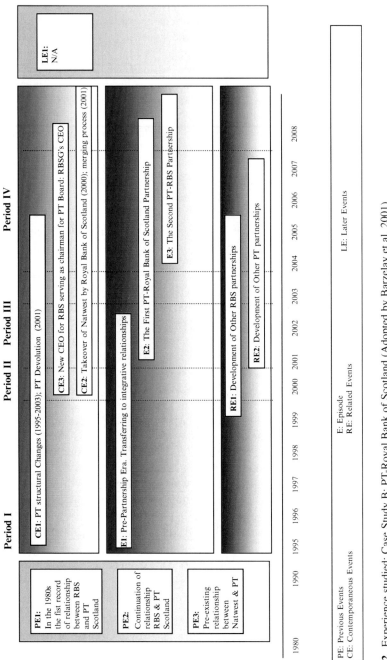

Fig. 3.2 Experience studied: Case Study B: PT-Royal Bank of Scotland (Adopted by Barzelay et al. 2001)

the formation of a partnership. They reflect the subjective perceptions of the interviewees. However, since there was agreement among the expressed motives by interviewees they are regarded as expressing the predominant organisational motives.

The organisations' motives are dynamic, meaning they change as a result of previous experiences, parallel to a person's motivation. According to Habermas, Mead, Durkheim and Freud it is through the socialisation process that most motives are shaped and as a result generate values and norms and the expectations of others (Mazur 2004) such as society, peer group, stakeholders and so on. As a result the motives of the organisations are sometimes mutually produced through symbolic interaction[21] with peer groups, partners and so on. Consequently an organisation may engage in actions without aiming to do so, or which it did not consider engaging in originally. However through interaction, an organisation's actions, or indeed the lack of them, acquire symbolic meaning within a set group of organisations. As a result an organisation may be forced to become involved in activities that are symbolically important to its peers, donors and so forth in order to abide by the predominant network culture. In this case the prevailing motive behind the action would be the need to be considered as part of a group which is dominating a sphere of interest, pertinent to the organisation.

Within a partnership there are two sets of motives in operation: the profit and nonprofit motives. Each organisation represents a different system of values and beliefs (Crane 1998; Stafford and Hartman 2000) due to their sectoral characteristics. The research question that this section will answer is: What are the motives behind the formation of NPO-BUS partnership? Are the motives shared between the two partners?

3.5.1 The Partnership Motives in the Earthwatch: Rio Tinto Case Study

The analysis of Earthwatch's motives suggests two dimensions. The first dimension is the *intention of the motives* manifesting the purpose of each motive. The intention of the motives, as indicated by the analysis, is divided into three categories of motivations: *instrumental (or extrinsic), idealistic and intrinsic motives.* The instrumental motives are guided primarily by the need of the organisation: (a) to safeguard its financial security; (b) to maintain its mandate as a non-confrontational NGO; and (c) to enhance the organisation's reputation and at the same time build Earthwatch's profile within the business community. Furthermore, the instrumental motives use the partnership as a means to an end. For example 'safeguarding financial security' can be achieved by attaining funding from a multitude of sources, such as members, individual donors, institutional donors, companies, government international funding bodies. In the case of

---

[21] Mead (1934, 1936, 1938), the founder of symbolic interactionism, saw interaction as creating and recreating the patterns and structures that bring society to life. The Symbolic Interaction of Mead was an early understanding of the interdependency between the different levels of reality.

Earthwatch, the associational form of the partnership is used as a vehicle to achieve financial sustainability for the organisation. The instrumentality lies in the importance that Earthwatch gave to this motive in order to move the relationship with Rio Tinto to another level, but also to attain know-how through the partnership in order to further acquire other corporate partners. The quote that follows is indicative of the instrumental intent of Earthwatch's motives:

> ...I mean the corporate sector for Earthwatch is of incredible importance, also funds. So ... funding is important. (Interviewee, Earthwatch)

Idealistic motives (i.e. changing the company) on the other hand derive from the idealism that surrounds NPOs, but are also an expression of the individual beliefs of Earthwatch's employees. The reason that I do not ascribe the idealistic motives to the organisational level is that there is: (a) no indication of idealistic motives in the mission statement of the organisation; (b) no indication of a planned method to achieve change within the company; and, (c) no system for monitoring change within the company. Hence, the motives that appear in the interview quotes related to change are classified as idealistic.

> ...And without that support and funding from the private sector we wouldn't be able to support those 140 projects in the field as well as we do now. So that's very important! But it's also very important for us that we feel that we can influence companies as well. (Interviewee, Earthwatch)

Finally, the intrinsic motives are related to the essential nature or character of Earthwatch as an NPO. Based on the analysis there are two basic intrinsic motives related to the organisation: (a) 'alignment of principles and values', and (b) 'educate and engage people':

> I think the other motivation is this thing about alignment of, if you like, principles and values and that we feel that RIO TINTO is a company that's committed to the environment, it's committed to improving its performance.... (Interviewee, Earthwatch)

> The motives will be about educating the employees at RIO TINTO in order to ensure more environmental responsibility on behalf of the company.... (Interviewee, Earthwatch)

The first dimension of the motives examined above was the intent of the motives. However, the analysis suggested another dimension to the motives. Organisations are continually evolving; their motivations are equally constantly changing in a dynamic fashion. However, motives related to both the organisational mandate and mission, or to the predominantly static sectoral characteristics of the organisation remain unchanged. Hence, the second dimension of motives, *the mode (state) of motives*, manifests in relation to whether motives remain unchanged due to their close relation to the organisational mandate and mission, or if they are dynamic to whether they have the potential of changing due to external circumstances. Hence, the mode of motives is manifested through two basic categories of motive: (a) the static motives; and (b) the dynamic motives. The static motives are characterised by their durability and strong relevance to the organisational mission. In other words, static motives are persistent within the organisation, they reflect the level of commitment to the organisational mission and hence they can be referred to as 'mission-led' motives. The dynamic motives are related to the operations of the organisation and they can be called 'organisation-led'. The dynamic motives change over time in order to adapt to internal or external circumstances, but would primarily leave the mission of the organisation unchanged. The mission-led motives might be influenced by the dynamic motives. In other words, the circumstances that an organisation faces might cause a change to its original motivation, i.e. its mission. However, in this case the organisation seriously needs to consider why this is happening; if and why it is the right direction for the organisation. The following passage demonstrates the two different modes of motives present, the static (mission-led) and dynamic (organisation-led) motives in Earthwatch:

> I think the motivation originally was that Earthwatch was a small and not very well-known NGO and it wanted to become larger and better-known for several reasons. One of them, for financial security, another because it was felt that enhancing the reputation and so on would build the profile and bring in more members and volunteers and so it would give more public support for Earthwatch's mission. (Interviewee, Earthwatch)

The static motives (mission-led motives) above are: (a) to achieve financial security; (b) to increase members and volunteers; (c) to ensure public support and the dynamic motives (organisation-led) are: (a) to build reputation; and (b) increase the size of the organisation.

As the previous and following passages demonstrate, organisational decision-making involves more than one motive being in operation at the same time:

> And again the motivation was ... partly mission-led and partly sort of organisation-led. So from the organisations' point of view it was obvious that if you had a large partnership with a company that perhaps had a five-year tenure and was financially larger, it was better for the organisation in terms of stability and long-term funding and so on. But also if you could get a major FTSE100 company to support you then that was like it would bring in others in similar types of partnership. (Interviewee, Earthwatch)

Based on all the interviews analysed and not only on the brief quotes presented above, a summary of both dimensions of motives – the intent of motives as well as the mode – related to the partnership is presented in Table 3.5.

The observations of Table 3.5 indicate that Earthwatch's motives are almost equally divided between static and dynamic. However, concerning the second dimension of the motives, it seems that there is a higher concentration of instrumental and idealistic motives. Since Earthwatch is a non-confrontational, non-political and non-campaigning NGO, having a majority of instrumental and idealistic motives seems to be in agreement with the organisation's characteristics. This finding is also in agreement with the research of Carmin and Basler (2002) who found that "EMOs engage in rational assessments and select tactics based on their evaluations of what will be effective in pursuing goals, but these evaluations are inherently influenced by identity" (Basler and Carmin 2002:B5).

Moving to the partnership motives for Rio Tinto it is interesting to look at an incident that contributed to the development of the partnership and which therefore sheds light on the motivation behind the partnership decision. An Earthwatch interviewee describes one of the critical incidents below:

> ... this may be an urban myth but the story goes that their Chairman went to a dinner party in London with a lot of staff from lower levels. And he was just sitting back listening to their conversations and there were all their partners there, and friends were there and that sort of thing. And they all... none of them said they worked for RIO TINTO. They were

Table 3.5 Earthwatch's two dimensions of partnership motives

	Intrinsic	Instrumental	Idealistic
Mission-led (Static)	Educate and engage people (employees)	Safeguard financial security	Make a difference within the business community
	Increase members and volunteers	Enhance organisation's mandate non-political, non-confrontational	Personal motives of Earthwatch employees (sub-culture)
		Ensure public support	
Organisation-led (Dynamic)	Alignment of principles and values with companies	Engage with any business with environmental impacts	Change/influence companies
		Build reputation within the business community	To influence companies by policy advising
		Make organisation larger	

all really ashamed working for RIO TINTO. And this is seven or eight years ago. And he thought 'My God! This is my staff who are ashamed to work for this company. Why is that?' And it's just transparent; it's all the environmental impact. So, he wanted to have a programme where ... well, he wanted to illustrate not only externally but internally with really key that the company was taking the environment seriously. And invest back in the employees. So, actually employee-volunteering prospect of Earthwatch alongside a non-confrontational aspect of serious science etc. and corporate experience but it seemed again a very good mix. (Interviewee, Earthwatch)

Rio Tinto faced image problems with both its internal and its external stakeholders. The primary motive for Rio Tinto to pursue a partnership with Earthwatch was to build a positive reputation for the company. Based on the data collected for this research (across all case studies, but also including the comparative interviews) it became apparent that the level of sophistication of a company's approach to CSR and its elements positively correlated to the extent and level of public pressure from its stakeholders. In other words, the higher the level of sophistication of a company's practices in CSR and its elements, such as community involvement, the higher the level the company was under attack in the past. That does not exclude the possibility of companies that were not exposed to reputational risk, or other non-financial risks, developing a higher level of sophistication in their approach.

As most criticism was arising from the radical environmental NGOs, it was imperative for Rio Tinto to address these issues first. As mentioned above, in 1995, the Friends of the Earth campaign against Rio Tinto was at its peak; also at the same time the Earthwatch interviewee makes reference above to the description of the 'dinner party' incident. The two events are identified as sources of occurrences within the case of the partnership episode. In other words, they contributed positively to the evolution of the relationship and hence enforced the motivation that is examined below.

In order to allow for a comparison between the motives of the two partner organisations a second level of analysis resulted in applying the same classification of motives that were used for Earthwatch's analysis, permitting consistency within the case study analysis.

Based on the eighth Annual Review of Rio Tinto's Social and Environmental performance of 2003, the following priorities were identified within the review:

(a) increase shareholder value; (b) build positive relationship with stakeholders; (c) increase corporate accountability; (d) retain transparency; (e) application of standards and controls; (e) uphold safety and occupational health standards; (f) finalise environmental standards; (g) enforce guidance document on applying human rights standards; (h) implement business integrity guidance; (i) implement compliance guidance.

The areas of priority within the company's Social and Environmental Report offer an indication of the level of sophistication in the company's CSR practices. They also can be regarded as part of the corporate mission of the company in the non-financial issues. As such I will use them in order to classify the motives into the different dimension and categories.

More specifically, the primary motivation for the company was to improve its reputation:

> Reputation. Essentially enhancing, managing and enhancing the reputation of the corporation. Reputation is in a sense everything because it is reputation that allows a company to operate. Particularly a resources company. For a resources company to continue to function, we need access to new sites, to develop new mines, that's putting it bluntly. If we had a poor reputation, then communities are likely to reject us. So it's very important for us that communities feel that we're an honourable company who will work in partnership with them to ensure that the benefits for that community and for that region are maximised from the development and through the life of a mine. (Interviewee, Rio Tinto)

It becomes apparent from the above quote that the primary motivation for the company was to improve its reputation, as this would allow the company to operate within both old and new communities, without facing opposition due to their previous record. In other words 'improving corporate reputation' is a static motive as it is related to the company's mission. It can also be classified as an instrumental motive since the aim is to employ a positive reputation in order to achieve or retain access to the old and new communities. Before suggesting a summary of the motives based on the analysis it is important to present another account from a different interviewee that demonstrates Rio Tinto's partnership logic and the shared opinions between different organisational actors across time:

> ...and we set up a structure whereby we basically wanted to improve corporate reputation and do that by improving our performance of the operations ... We looked into the corporate reputation and HR Department, looking at the technical aspects of biodiversity management. And so we started to look at developing a biodiversity policy and strategy and, at the same time, we looked at the possibility of setting up a triangle, a suite of partnerships with organizations, which could help us develop our policy. ... So, we decided we try and look at the concept of partnerships, to see whether it was a valid concept, to see whether it worked. (Interviewee, Rio Tinto)

As both accounts are placing the "improvement of the corporate reputation" as a clear priority in the company's motivations, it offers a validation of the level to which this particular motive is embedded within the company. These two accounts also offer a route to tracking down the company's partnership logic.

Moreover the company's motives include to "gain access to NGO expertise", which is classified as a static motive since it is manifested in the company's mission (see above: finalising environmental standards, build positive relationship with stakeholders). To "improve performance of operations" is classified as a dynamic motive, since it is 'organisation-led'. Also it is considered an instrumental motive as it is guided by the need of the company to improve its corporate reputation. Furthermore the motive to "develop a biodiversity policy and strategy" is regarded as an intrinsic motive (reference in the company's mission: application of standards and controls) and as dynamic since it relates to the operations of the organisation rather than its mission.

Finally, the aim to "demonstrate internally (internal stakeholders) and externally (external stakeholders) the importance of the company's environmental priorities" (see quote on critical incident above) is classified as a dynamic motive since it is related to the organisation's ongoing operations and is dependent on the dynamic external environment (different pressures from stakeholders). Also it is regarded as an instrumental motive since it is being achieved primarily through the company's partnership with Earthwatch.

As is apparent from Table 3.6 Rio Tinto's interviewees did not mention any partnership motives that could be classified as idealistic. This was expected as: (a) this is an organisation that belongs to the profit sector whose corporate culture does not allow idealism; (b) Rio Tinto is a particularly sophisticated company and even in the case of community involvement its strategy is highly influenced by the priority of the financial bottom line. The expressed motivation is primarily due to the early pressure the company was facing which was ultimately to its own advantage as it developed organisational capabilities that place it today as a leader in community involvement, even beyond its own sector (based on remarks by other organisations about Rio Tinto).

Rio Tinto's motives are almost equally divided between intrinsic and instrumental and static and dynamic. There is a slight concentration on instrumental motives and particularly on dynamic motives. However, the difference is not considered important. The motives are equally divided between the categories, which can be seen as an equal interest in addressing 'image' as well as 'operations' issues. This is in accordance with the company's rhetoric around 'embracing change', 'being committed to responsible practices', but on the other hand with the company's critics that Rio Tinto is interested in 'greenwash' as it is interested in improving its reputation. However, it offers testimony that in order to retain its sustainability, a

Table 3.6 Rio Tinto's two dimensions of partnership motives

	Intrinsic	Instrumental	Idealistic
Static/generic mission-led	Gain access to NGO expertise	Improve corporate reputation	N/P
Dynamic organisation-led	Develop a biodiversity policy and strategy	Improve operations performance Demonstrate internally (internal stakeholders) and externally (external stakeholders) the importance of the company's environmental priorities	N/P

N/P: Not present

clear transition from the old corporate line[22] to a new more transparent and open company needs to be retained.

The next section will examine the motives for both partners and attempt to compare their respective motivations in order to examine the extent to which motives were shared among the partners.

3.5.2   Earthwatch-Rio Tinto Motives Compared

The comparison of the business and the nonprofit motives aims to discuss the extent to which the motivation is shared between the two partners.

The first observation is that the organisational motives are compatible.[23] By compatible I mean they form a homogenous relationship that would not pose any threats or challenges to either partner. In other words, based on the definition of compatibility, the two organisations' motives allowed for the formation of a homogeneous relationship where the interaction among them would have a limited scope for change. Rio Tinto chose a partner whose organisational characteristics and identity were non-threatening, aiming to get a policy input from an NPO in an area that radical parts of the environmental movement were contributing to the increase of primary[24] risk through secondary risk for the company. Earthwatch's non-threatening motives allowed the company to adopt a step-by-step approach towards NPOs by building its reputation within the sector, while also allowing it to establish similar relationships. On the other hand, Rio Tinto's motives permitted Earthwatch to gain experience in forming partnerships and ultimately to build its reputation within the business community. The compatibility of motives also meant that the process of partnership building and delivery of their objectives was going to have minimal risks from within the partnership. In other words, the motives' compatibility with the organisational characteristics safeguarded the delivery of expectations and minimised potential risks from within the relationship.

Earthwatch's mission-led motives include broader motives that concern the environment and society; however they appear to be idealistic in nature rather than instrumental. In other words, although Earthwatch was interested in achieving alterations within the company there is no evidence of a strategic and evaluative intent. On the other hand, Rio Tinto's motives are exclusively centred on its own circumstances and needs. Only in the intrinsic and dynamic motives does the biodiversity policy and strategy imply a broader interest in natural capital, which equally results in benefits for the environment and society.

[22] Heap (2000) also agrees that "Rio Tinto has made efforts towards genuine engagement with NGOs" as early as 2000.

[23] According to the Merriam Webster Dictionary one of the meanings of being compatible is "capable of forming a homogeneous mixture that neither separates nor is altered by chemical interaction" (Webster 2004). In other words when two things are compatible there can never be an alteration by interaction. This is a concept that this research explores in the discussion chapter.

[24] Primary risk according to Power (2004) refers to the direct material or financial risk and secondary risk refers to reputational risks.

Also, as noted earlier there were many motives in operation at the same time in both organisations. Since, this research is exploratory in character there is no evidence with regard to the priority of each motive in operation.

It is understood that the above motives are not integrated within the company's broader motivations, as this is not the purpose of this study. However, I recognise that as a limitation of the current analysis.

3.5.3   The Partnership Motives in The Prince's Trust-Royal Bank of Scotland Case Study

Following the categorisation of the previous chapter and as indicated by the analysis, the *intention of the motives* is divided in three categories of motivations: *instrumental, idealistic* and *intrinsic* motives. The instrumental motives are guided primarily by the need of the organisation to: (a) safeguard its financial security by (i) off-setting core funding expenditure through the partnership and (ii) funding change and innovation; (b) improve the brand name of the organisation and its reputation by establishing a partnership with a 'blue-chip' company; and (c) enhance the organisation's competencies by utilising and adopting the bank's approach, thinking and ideas in order to allow for change to take place. With regard to the instrumentality of the partnership motives, PT uses the partnership as a mean to safeguard financial security and sustainability. Funding is still important for PT, however attaining resources from the profit sector appears more balanced from a multitude of resources (50% from government, 13% from the profit sector and 37% from other sources). The quotes that follow are indicative of the instrumental intent of motives:

> …Everybody thinks that because The Trust is driven by the Prince that it must be a very lucrative charity and it's not. So, the financial drivers are very important. (Interviewee, PT)

> … and that was something else we needed to do through this approach was to ensure that things that we were writing into the budget for this funding were going to not necessarily create new expenditure but offset core planned expenditure and … in a different direction..... (Interviewee, PT)

> …there was a real motivation … to get their ideas, their influence in terms of ways of thinking, ways of approaching, ways of problem solving … … … I suppose there was also a motivation about … being able to say we've got such a big partnership. … Big in terms of money, big in terms of scale and sophistication. (Interviewee, PT)

> Another key driver for The Trust is getting a good brand name. So if you're going to have a quality blue chip organisation standing behind you … the Royal Bank Group is a big networked organisation. (Interviewee, PT)

Finally, a decisive motive for The Trust was the fact that RBSG posed a serious creative question to its potential partner:

> "Where would you like to be in three years?" (Interviewee, PT)

This question allowed PT to experiment with its core programmes, undergo a process of change and adapt a much-needed new approach:

> as part of that we recognised we had to change the way we worked and therefore we needed a partner that enables us to give us the freedom to try these things and change the way we worked to reach those people.… (Interviewee, PT)

And also because the nature of the way in which the RBS gave us the opportunity to go into partnership. They asked us where we would like to be in three years, which I think got everybody thinking here and certainly that changed, you know, our perception of what we could do with a commercial partner. (Interviewee, PT)

The meaning of the idealistic motives (i.e. in the Earthwatch-Rio Tinto case study, i.e. changing the company) was not present in the case of PT. There was no evidence, either on the organisational level or on a personal level (i.e. references to personal motives of employees) that there was an intention to change the business partner or influence its practices. On the contrary this case study appears to be diametrically opposed to the previous one on this particular issue: in the first case study the motivation was placed within the nonprofit sector aiming to deliver change to the profit. In the second case study it appears that the change motive appeared in the interviews of profit sector i.e. to facilitate change within its nonprofit partner.

Finally, the intrinsic motives relate to PT's organisational characteristics. The three basic intrinsic motives of the organisation are: (a) to meet the social inclusion objective of the organisation; (b) to increase volunteers by establishing a relationship with a network organisation (such as the RBSG); and (c) setting up a blueprint partnership, which would assist in developing similar partnerships not only with other businesses but also with the government.

So I think that we would as an organization have ticked their boxes and they would have ticked our boxes simply in terms of their social inclusion objectives and the work that we do ... Because we want their employees involved.... (Interviewee, PT)

The above quote not only mentions the central place of the 'social inclusion objectives' for both organisations, it also illustrates the perception of 'perfect fit' between the two organisations. The above refers to the focus of PT work and the priority of the 'social inclusion' theme within the bank's CSR agenda. As suggested by the interviewees, it is the background of the individuals involved in the community investment department and their experience as government and nonprofit sector employees which allowed them to develop a better understanding of both the priorities of the government agenda and the difficulties charities face in the implementation of their objectives.

PT is a volunteers' organisation, as earlier quotes mentioned above, hence increasing the intake of volunteers from the RBSG was a priority. Furthermore, a set of special roles were developed in The Trust due to the partnership, such as: "Employee Involvement Account Manager for the RBS Group". This, in my opinion, highlights the importance of this intrinsic motive for PT. The quote that follows from a RBSG employee is also in agreement with this observation:

And, I think the PT finds bank staff, they're really interested in having bank staff as volunteers and aftercare advisors because most people in the banking world will have knowledge to help young businesses. (Interviewee, RBSG)

According to the conceptualisation applied in the first empirical chapter, the second dimension of the motives refers to the *mode of the motives*, manifesting whether the motives remained unchanged due to their close identification with the organisational mandate or if they have altered due to external circumstances. Hence, the two categories of the mode of motives are: (a) the static motives and (b) the dynamic motives.

The generic motives or mission-led motives for PT are: (a) to achieve financial security; (b) to increase the organisation's volunteers; and (c) to meet the social inclusion objective of the organisation. The dynamic or organisation-led motives are: (a) to set up a blueprint partnership, which would assist in developing a similar partnership; (b) to improve the brand name of the organisation and its reputation by establishing a partnership with a 'blue-chip' company; and (c) to enhance the organisation's competencies by utilising and adopting the bank's approach, thinking and ideas in order to allow change to take place.

As we can observe, multiple motives are in operation in the organisational decision-making, as was similarly the case with the previous partnership analysis. Based on all the interviews analysed for PT a summary of both dimensions of motives is presented above.

Based on Table 3.7 it appears that PT motives are almost equally divided between static and dynamic. However, concerning the second dimension, it seems that there is a marginally higher concentration of instrumental motives. An important observation is the lack of idealistic motives as they have been defined above, contrary to the existence of idealistic motives for Earthwatch. Although there were remarks, as the quote below suggests, hinting at idealistic motives, these were not manifested within the organisational mandate, either at the beginning, or throughout the first 3 years of the partnership:

> Now that's something where actually the bank might take an even greater role and where we might influence the way they operate in a certain part of their organisation. (Interviewee, PT)

Table 3.7 PT two dimensions of partnership motives

	Intrinsic	Instrumental	Idealistic
Static/generic (mission-led)	Meet the social inclusion objective of the organisation Increase volunteers	Safeguard financial security: 1. Off-setting core funding expenditure through the partnership 2. Funding change and innovation	N/P
Dynamic (organisation-led)	Setting up a blueprint partnership, which would assist in developing similar partnership	Improve the brand name of the organisation and its reputation by establishing a partnership with a 'blue-chip' bank Enhance the organisation's competencies by utilising and adopting the bank's approach, thinking and ideas in order to allow change to take place	N/P

N/P: Not present

A partial explanation of the absence of idealistic motives could be that social charities, such as PT, specialising in youth, do not have the experience and sophistication of environmental NPOs. In the case of the environmental movement, that challenged businesses systematically for their lack of socially responsible environmental practices, influencing the change in business practices is quite high in the motives for partnership formation. In the case of PT, being a non-confrontational and non-campaigning charity, having a significant majority of instrumental motives seems to be in agreement with the organisation's characteristics. However, the absence of idealistic motives indicates that possibly the charity does not yet recognise its potential to facilitate change in the profit sector.

Moving on to the motives for the RBSG, this section examines the reasons behind the formation of the partnership from the Royal Bank of Scotland's perspective.

The Royal Bank of Scotland did not have any pressing image problem with its stakeholders (as was the case with Rio Tinto), apart from the criticisms previously presented in the section on organisational characteristics: 'excessive' profits, executives' level of bonuses, credit card debt and loan problems. It is interesting that I did not encounter any criticism referring to the risk assessment method that banks use before they decide to grant start-up loans and which prohibits access to young people from disadvantaged backgrounds, people who PT supports. Interestingly this issue is considered to be merely part of the reality which disadvantaged young people have to face. However, interestingly this issue is precisely the interface between the bank and The Trust. More importantly the bank's community investment with its CSR agenda is focused around this issue and it is the area in which PT can make a difference within the marketplace. The PT appears hesitant to play this new role. A new set of responsibilities would arise from this role which would be beneficial not only to the PT but also to society at large.

The 'employees' and the 'community' are the two priority stakeholder groups for the community investment department of the RBSG. Based on the interviews as well as on the documentary evidence collected, the priorities for the bank's CSR agenda are:

> Our Group activities are focussed on education and promoting social inclusion. We achieve this by developing long-term partnerships with charities and Government. At an individual level, we believe that our people are better placed than we are to identify the needs of the communities in which they live and work. For these reasons we support a broad range of community-based causes selected by our employees. (RBSG 2004a:40)

Since social inclusion was a priority for the bank, PT was selected as a 'natural' partner for RBSG in order to meet its CSR agenda. Although the relationship between the two partners existed previously, it was the proactive approach of the bank that resulted in the partnership formation, as with the rest of the bank's relationships with the nonprofit sector. It is important to note that when a project is introduced from the top of the organisation there is a greater propensity for longevity (Austin 2000), however the survival of a project is dependent on the bank's staff interest and support for the particular project.

Well, there's a number of drivers for any business I think, key drivers around CSR agenda obviously our first and foremost it's a good thing to do, but there are business benefits around staff moral, staff retention, staff recruitment, potential business benefits and for the company as well as benefits for the partner organisation and also societal benefits. And ultimately as an organisation we're successful because the communities in which we operate are successful. So, it's also we're part of that community, we're not a separate, stand-alone kind of... we're actually in the heart of those communities that we're supporting and it's important that they do well and as a result we do well. (Interviewee, RBSG)

As it becomes evident from the above, the expected business benefits functioned as key drivers that encouraged the partnership relationship. The volunteering opportunities the partnership offers to the bank function as a Human Resources (HR) development opportunity that assists with staff retention and motivation, since they are matched against the RBSG's core competency framework which benefits the careers of the staff at the same time. Another stakeholder group that has been prioritised in the bank's analysis is the government:

... there was a need to in the eyes of the government to be seen to be more strategic in their approach to their community investment and their community involvement .... Agenda. (Interviewee, PT)

Well because we have a socialist government as an expectation that we will go and support communities within the environment that we're currently in. ... We did it when we had our Tory government, the same sort of commitment... bearing in mind as well we don't in this country have stringent community tax laws as there are present in the States, and the community reinvestment tax ... ... is not ... here. I think we all prefer that to be the case, so one way to avoid the need for any form of taxation to support community is actually to be seen to be ... those communities at the outset. (Interviewee, RBSG)

The regulatory power of the government operates as a driver for the bank to become proactive in its CSR strategy by forging partnerships with the nonprofit sector, hence lowering the risk for any form of new taxation and at the same time taking advantage of the existing tax benefits.

Further motives that resulted in partnership formation were: the publicity benefits, the enhancement of the bank's community role, the brand association with the PT, developing new and second generation business in the community and the fact that PT was a UK-wide charity that could offer HR opportunities in all RBSG branches. As can be seen from the existence of idealistic motives, philanthropy is still in operation within the bank, summarised in the statement: "it is a good thing to do." Similarly idealistic motivation appears with reference to the 'societal benefits' in the interviews but also in the documentary evidence at hand.

Last, but not least, an important motive for the bank was to facilitate change in The Trust:

...the primary objective was we believed that our cash injection would help the PT achieve a shift in the way it did things... the bank was looking for the right partner to do something with young people to help them into education, employment and enterprise which was our primary theme.... (Interviewee, RBSG)

Table 3.8 RBSG's two dimensions of partnership motives

	Intrinsic	Instrumental	Idealistic
Mission-led (static)	Cover the whole of the UK regions Develop new and second-generation business	Tax benefits Meet the social inclusion objectives Increase HR development opportunities	Do the right thing (sub-culture)
Organisation-led (dynamic)	Setting up a blueprint partnership, which would assist in developing similar partnership Retain the communities' and bank's sustainability	Brand association enhancement Publicity Demonstrate to the external stakeholders, government and local communities the bank's role in the community Facilitate change within the nonprofit organisation	Achieve social benefits

Facilitating change meant fostering a proactive approach and assisting The Trust to become more effective in its mission by allowing it to experiment and develop a new approach to its programmes, but at the same time to be able to evaluate the programme outcomes:

> I mean there's volume of people helped, there's outcomes for young people to measure and evaluate. You know, we look at … we need to start driving more and more onto cost effectiveness. (Interviewee, RBSG)

Based on the above but also in the analysis of all the interviews and the documentary evidence, the RBSG's motives for establishing the partnership with PT are categorised and presented in Table 3.8.

The reason I have included the motive 'facilitate change within the nonprofit sector organisation' in the instrumental motives (unlike the Earthwatch motives related to change) is due to the existence of a strategy on behalf of the bank. On the other hand, and unlike Rio Tinto, the interviewees at RBSG did mention partnership motives that are classified as idealistic. Although this was not expected, a plausible explanation is the non-traditional banking multisector background of the employees.

RBSG's motives are almost equally divided between intrinsic and instrumental and generic/static and dynamic, accompanied by the existence of idealistic motives. There is a concentration on instrumental motives and particularly in dynamic motives. However, the difference is not considered significant. The motives are equally divided between the categories, which can be seen as an equal interest in addressing issues surrounding 'image' as well as the community investment goals, however not changing their loan practices with regard to disadvantaged young people. This is in accordance with the company's rhetoric around 'embracing change', 'being committed to responsible practices', but on the other hand with the need to address image issues due to the importance of the CSR agenda in the way businesses operate in the UK, particularly when they need to be seen as advancing the CSR agenda according to the government's priorities. In the case of the RBSG there was no need for an identity shift for the organisation, however since it recognises the importance of the CSR agenda it plays an important part in furthering the debate which adds to its own reputation within a number of stakeholder groups and hence safeguards its own sustainability.

The next section will examine the motives for both partners and attempt to compare the motivations in order to examine the extent to which the motives were shared among the partners.

3.5.4 Motives of the Prince's Trust-Royal Bank of Scotland Compared

The comparison of the business and the nonprofit motives aims to discuss the extent to which the motivation was shared between the two partners.

It can be observed that the organisational motives are compatible between the two partners, as was the case in the previous case study. Based on their organisational identity, neither partner poses any threat or challenge to the other. The motivations of RBSG, the partner that initiated this partnership, were primarily: (1) meeting its community investment mission and CSR agenda; (2) improving its brand image; and (3) increase the volunteering opportunities. Hence, according to the perceptions of the interviewees, the partners motives' instrumentality along with their mission-led motives present a 'natural fit' scenario for both of them. It is interesting that the reason the bank is motivated to facilitate change in the charity is linked with its mission-led motive of social exclusion, which in turn corresponds with the mission-led motive of PT. The roots of the relationship go back 20 years and it seems that the latest merger between RBSG and Natwest resulted in a further proximity between the partners allowing for a shift in the organisational form of the relationship to a higher, more integrative, stage relationship.

The compatibility of motives also meant that the process of partnership building and delivery of the objectives was going to have minimal risks from within the partnership. In other words, the motives' compatibility safeguarded the delivery of expectations and minimised of potential risks within the partnership.

Also, as noted earlier, there were many motives in operation at the same time in both organisations. This research has no evidence with regard to the priority of each motive in operation.

The brand association and a number of commonalities across the two organisations which constitute the reasons for the initiation of this partnership, included: (1) royal affiliation; (2) UK-wide organisations; (3) social exclusion being strongly identified with both organisations; (4) interest in business start-ups; (5) sharing the same goal for the partnership: to reach 550 volunteers from RBSG.

> ... I mean brand associations obviously are important. I think we've got some common objectives through the bank's Community Investment programme and our programme and

what we're trying to achieve and obviously some commonalities which is why this partnership exists in the first place anyway. (Interviewee, PT)

The above motives are not integrated within the company's broader motivations, as this is not the purpose of this study.

3.6   Opening Pandora's Box: Familiarity, Compatibility and Synagonistic Relationships Across the Sectors

This section discusses the above findings in light of the 35 comparative interviews that were conducted within 29 organisations in order to confirm, disconfirm or extend the findings of the formation phase of partnerships.

Earthwatch and The Prince's Trust represent two different NPOs within each case study. The contrasting organisational characteristics include differences in the industry sector, size, revenues, mode, scope of operation and reputation.

They are both similar with regard to the mode of approach towards businesses hence they are both collaborative NPOs. As suggested by the research of Berger et al. (2004) increasingly NPOs have inherent in their missions a collaborative approach towards business establishing close relationships, including partnerships.

This was not only the case with Earthwatch and The Prince's Trust but also with other NPOs that participated in this research. For example, the interviews with A&B (previously known as ABSA[25]), the CBI (Confederation of British Industry), WBCSD (World Business Council for Sustainable Development), WWF and IPRA (International Public Relations Association) demonstrated an inherent collaborative approach towards BUSs. The above NPOs were traditionally positive towards the profit sector; however, others, such as Greenpeace, appear to have similarly embraced more collaborative approaches, even if only selectively with some companies. Greenpeace has developed a non-financial[26] partnership with NPower, an energy company in order to provide alternative forms of power to British households. Even Friends of the Earth which opposes currently partnerships, already work on a small scale with companies as they need to address the profit sector's prominence in the global debates. Involving BUSs in their agendas seems to be an imperative in order for NPOs to avoid 'issue isolation' by broadening their agendas and remaining relevant to the global debates and hence increase their impact, as the following quote from Greenpeace demonstrates:

And I mean the thing that I was saying about Greenpeace is our agenda is very narrow. I mean we're not a sole-issue [NGO] but we're very much focused on environmental issues. I have to say as an organisation we're starting to broaden that now because, you know, issues of environment and development often go hand in hand ... moving forward, we need to be very mindful of other issues that affect environmental issues like trade, globalisation and things like that. ... the more international we become the more the international issues affect

---

[25] A&B is one of the most significant UK organisations in the Arts.

[26] No financial support for Greenpeace comes from NPower.

what we do really. ... I mean if we do not take account of some of the issues that surround
that, we will be left behind basically, and also irrelevant. (Interviewee, Greenpeace)

As indicated by the interviews a number of NPOs still remain confrontational
towards BUSs, such as the Partizans (a sole issue NGO) that uses a network structure.
Partizans suggest that they criticise partnerships between collaborative NPOs and
BUS on the grounds of lack of in-depth knowledge, for example, of the mining
industry:

So you have these NGOs that actually don't know very much about mining. They're visiting the projects, they're saying 'well that looks good', and we've seen this study and they
appear to have done this environmental impact assessment and yes, we can put the rubber
stamp on it. I mean, to me, that's, that's, it's unacceptable. (Interviewee, Partizans)

If indeed a partnership was going to take place between Partizans and Rio Tinto
then the claim of historical adversarial relationships portrayed in the literature
would stand. However all partnership relationships examined in this research stem
from historically synagonistic[27] relationships. Hence a high degree of familiarity
existed among the partners. This became evident through the historical dimension
of the relationship within the two case studies and was further verified through
comparative interviews. A variation, for example, was presented in the partnership
between WWF and Aviva. The two organisations did not have institutional relationships, but previous collaborations existed between individuals that were carried through to their next job. Hence, the previous collaboration functioned as an
institutionalisation process on the micro/personal level that allowed for a rapid
organisational institutionalisation despite the lack of historical organisational
relationship.

Hence, the formation of a partnership relationship between collaborative NPOs
and high- or medium-impact BUSs (according to the FTSE4GOOD classification)
stems from synagonistic previous interactions, either on the personal or organisational level, that result in more intense and frequent relationship that Austin (2000)
classifies as integrative, where partnerships are positioned. However, based on the
interviews, the organisational actors did not separate clearly the different *forms of
association* such as sponsorship, partnership and the *approaches of interaction* such
as transactional approach or partnership approach. This will be further discussed in
the final chapter of the book.

The interviewees indicated a clear shift within the focus and language of
interaction from sponsorship to partnership. According to the information officer
at A&B it was the CSR discourse that posed the challenge for a multi-level and
multi-purpose interaction between the arts and businesses. The interviewee
suggested that the shift started in 1999 and reached its peak in 2001 when the
European Union's Green Paper appeared as a result of the consultation process,
which was characterised as 'Pandora's Box'. Another driver for the shift at A&B
was the government's discourse which gave prominence to concepts such as

---

[27] The opposed of antagonistic relations is synagonistic relations (Miller and Stephen 1966).

'creativity, innovation, CSR and social inclusion' as clear priorities for the government's agenda. However, as the interview revealed these concepts are usually viewed by practitioners as 'buzzwords', rather as clearly defined concepts that are rooted in organisational, cultural or mindset changes. Hence, quick changes that are not embedded within organisations are associated initially with rhetoric rather than representing a need that stems from the organisation's reality. The above justifies the confusion between the forms and approaches of association and the lack of conceptual clarity through the use of partnership language.

With regard to the motives, two dimensions have been identified through the analysis of the intent and mode of motives. The intention reveals if there is a linkage between the reasons behind the partnership formation and a strategy that will lead to outcomes. In the case of NPOs it appears that the motives associated with a strategic intent within both Earthwatch and The Prince's Trust are 'safeguarding financial security', 'building reputation and enhancing the brand name' for the organisation. This is supported by the rest of the interviews with the exception of the WWF-Aviva partnership. The relationship did not involve any monetary resources being exchanged and it was predominantly an informal exploration into the understanding of sustainability between individuals across the two organisations for the financial sector industry in order to improve its practices. Hence 'safeguarding financial security' did not constitute a motive for WWF. However, reputation either formally (in the case of publicised partnership) or informally (known only to specific interested publics) appears to be a potent motive for engaging in interaction with well-known organisations. Other instrumental motives include improving organisational competencies through the inherent capabilities of the partners. The societal outcomes are indirectly presented within idealistic motives in the case of Earthwatch, since there was no indication of strategic or evaluative intent. In the case of PT there was no indication of motives related to societal outcomes as opposed to social outcomes which are present in both organisations within the intrinsic motives.

The second dimension of motives, being static or dynamic, demonstrates the inherent link between the organisational mission and partnership as a strategy for achieving the mission of collaborative NPOs. The dynamic motives emphasise the reasons for a partnership relationship with BUSs and the static motives the internal needs of the organisations. Hence it appears that *partnerships are a strategy for achieving organisational outcomes based on the internal and external needs.* In some cases the motives might reveal conflict between the organisational mission and partnerships with BUSs as a strategy for achieving organisational outcomes as opposed to societal outcomes. This issue will be further discussed in the last chapter of the book.

In the case of the BUS partners the intent of the motives prioritises the instrumental motives that are linked with using the partnership for reputational reasons, publicity and association with the nonprofit sector. In the case of RBSG one of the motives is to facilitate change, therefore assisting improvements within the partner. Both BUSs view the partnership relationship as a way to improve their capabilities within the NPOs' areas of specialism (develop a biodiversity policy, increase HR opportunities and meet the social inclusion objectives). These findings were confirmed by the comparative interviews with the exception of the WWF-Aviva partnership as explained above. Corporations either reactively (in the case of Rio Tinto) or proactively (in the case of RBSG) form partnerships with NPOs in order to address their internal demands that are associated with CSR and in particular community involvement or environmental issues. These issues become a priority gradually through the criticism they receive based on their practices and operations which pose threats to a number of stakeholders. The interviewee from BAA Company, who was interviewed, epitomised this with the following statement: "trouble is our friend". This is further confirmed by the following quote from BP:

> ... we've had our fingers burnt by being arrogant in the past and I hope that we are changing our ways. I think, yeah, we learnt in Colombia about ten years ago that you can't ignore the community and you can't ignore the way you do business that affects that community. And at the end of the day you'll just have higher costs and higher security bills and disrupted business if you don't understand what the community's concerns are. ...Absolutely! [the above acts as motivation for change] And I think that, you know, there's a growing understanding of that piece of our impact on society and its reaction to that. (Interviewee, BP)

The comparison of the partners' motives revealed a high degree of *compatibility*. In the case of collaborative NPOs having partnerships with BUSs, motives are driven by *anticipated organisational outcomes* rather than conflicting agendas being met between organisations with different understandings on issues of mutual interest.

The above findings are further discussed within the last chapter of the book in order to develop conceptualisations that encompass all three partnership stages under examination.

3.7   Conclusion

The first empirical chapter concentrated on the formation of partnerships offering an analytical framework in order to analyse the formation stage of partnerships by examining: (a) the organisational characteristics, (b) the historical dimension of the relationships, and (c) the motives of the partner organisations. The last section of the chapter discussed the findings in light of the comparative interviews across 29 organisations that participated in the research.

One of the main findings of this chapter is that the predominant organisational characteristic associated with NPOs is their mode of approach towards BUS. They are defined as collaborative NPOs and demonstrate an inherent collaborative approach towards the profit sector. The historical dimension of the relationship between collaborative NPOs and medium- or high-impact BUSs appears to be synagonistic as it does not reveal any conflict that would be evident through the organisational characteristics, mission, or motives for developing a partnership relationship. The motives between the partner organisations appear compatible, driven primarily by anticipated organisational outcomes that originally bring the two partners together rather than addressing a social issue from a different, and at times conflicting, perspective.

The next empirical chapter discusses the implementation phase of the partnership relationship within the case studies and the comparative interviews.

Chapter 4
Stage Two: Partnership Implementation

4.1 Introduction

The implementation is the second stage of partnerships within which the interactions between the partners are discussed. This chapter of the book looks into the phases of implementation across the Earthwatch-Rio Tinto and The Prince's Trust-Royal Bank of Scotland case studies. It identifies the processes that take place and discusses the dynamic exchanges across the partner organisations. The final section of the chapter discusses the findings of the two case studies within the implementation stage and examines if conflict is observable between the partners in light of the comparative interviews that were conducted for the research across 29 organisations.

4.2 Partnership Phases in the Earthwatch-Rio Tinto Case Study

The first part of the chapter presents the interactions within the first case study. The different phases that are identified within the Earthwatch-Rio Tinto case are: partnership selection, partnership design and partnership institutionalisation and are presented below.

4.2.1 Partnership Selection

The first stage of the relationship's evolution is the *partnership selection*, which started when Rio Tinto made the decision to develop partnerships, rather than other associational forms, in order to achieve the company's strategic objectives (i.e. "improve corporate reputation by improving operations performance"). It was a conscious decision that aimed at testing out the partnership concept "a slightly sexy subject at the time" (Interviewee, Rio Tinto). As Rio Tinto was interested in developing not one but rather a range of partnerships, assessing the options of possible partners involved talking to various nonprofit organisations in order to determine the potential of each option:

M.M. Seitanidi, *The Politics of Partnerships: A Critical Examination*
of Nonprofit-Business Partnerships, DOI 10.1007/978-90-481-8547-4_4,
© Springer Science + Business Media B.V. 2010

I spent a lot of time over a period of some years discussing RIO TINTO and issues surrounding it in the mining industry with environmental organisations, particularly in the UK, so we knew quite a lot of environmental organisations quite well. So in some cases it wasn't necessary the best way of setting up partnerships, but it made sense to try and develop the first partnerships with the organisations that we knew very well, like Earthwatch, so that's what we did. (Interviewee, Rio Tinto)

The deliberation of the partnership meant that it was part of a planned rather than an emergent process and it was based on a number of criteria. The level of familiarity between the two organisations was an important factor as explained in the quote above. According to Earthwatch interviewees among the reasons that Rio Tinto chose the environmental nonprofit organisation were: (a) Earthwatch had a good track record with other companies; (b) both organisations were working in the same geographical areas around the world and in particular they both had headquarters in Australia and the UK; (c) Earthwatch was cost-effective regarding the ratio money/positive exposure it offered to the company; (d) Earthwatch was a professional nonprofit organisation; (e) Earthwatch provided a safe profiling platform for the company; (f) both organisations were operating within similar time-scales; and (g) the personal 'chemistry' between the people worked effectively. On the other hand Earthwatch's decision was primarily based on the fact that Rio Tinto was a company committed to change, as the quote testifies:

So that's the basis on which we work with the companies that we believe that they are committed to improving not that they are necessarily good or bad. (Interviewee, Earthwatch)

However, arriving at the above conclusion is by itself a very difficult[1] task:

... and is very difficult for small NGOs, we can't rigorously assess a large company who's operating around the world but we can look at indicators in terms of how they're reporting in their social and environmental report, how do they perform on FTSE4GOOD and see if it's good, how do they perform on with the BIE (Business in the Environment) index for example. We have our one-on-one relationships and our one-on-one meetings where we can ask them. Yes, so ... if you like, there are a range of criteria that we can use at a certain level in terms of understanding their policies and procedures and what they're reporting I guess and what targets they set and that sort of thing. But most of it boils down to probably our individual relationship with a company, which means our individual relationship with probably one or two people and the feeling we get from those one or two people about their own commitment. (Interviewee, Earthwatch)

Also, Earthwatch used another measure as a risk assessment process:

The other measure is ... is there a company policy on environmental issues or on biodiversity and is there a designated person on the board responsible for it? And is that policy endorsed by the board chairman? So you're looking for that high level commitment and then you're also looking for the sort of systems or procedures within the company that demonstrate that that high level commitment is being delivered. (Interviewee, Earthwatch)

---

[1]ERM is the consultancy that verified the quality of Rio Tinto's data presented in their social and environmental report. The consultancy stated that it is impossible to assess all operations of Rio Tinto. Hence a similar task would be impossible for a small NPO such as Earthwatch.

As it becomes apparent from the above quotes there are two ways in which Earthwatch assessed the risk of partnering with the company: *a formal and an informal method*. For Earthwatch, the formal method appears to comprise of looking: (1) at corporate indicators in terms of how they are reporting in their social and environmental reports; (2) at the FTSE4GOOD indicators; (3) at the BIE[2] index; (4) at the company's policies in place; (5) at whether there is a board-level commitment regarding environmental policies. According to an interviewee, at the time Earthwatch was in the middle of developing a policy on how they would work with companies.

Looking at the indicators as a part of a formal risk assessment method for the pre-partnership stage, as suggested, can be an equally ambiguous process. For example, in the case of Rio Tinto, FTSE4GOOD did not include the company in its index based on the company's uranium production, hence there were no indicators. On the other hand, the BIE Index is a voluntary self-assessment process of the company, which does not guarantee a third-party assessment.

Although the formal risk assessment method for Earthwatch was briefly described during the interview it appears that it was mainly the informal method that was in operation at the pre-partnership stage of the Earthwatch-Rio Tinto relationship. The informal risk assessment method for Earthwatch seemed to comprise of one-to-one meetings with the company's executives, usually one or two people within the company. The meetings are likely to assess the commitment towards change of these particular corporate executives.

A limitation of the informal method is the small number of individuals that are used as a way to assess a multinational's commitment to biodiversity and hence to changing their policies and practises. Furthermore, within the informal risk assessment method an important aspect of the decision-making process within Earthwatch was the micro-political process that occurred within the organisation. Earthwatch employees raised their concerns with regard to the decision to form a close relationship with Rio Tinto:

> … there was certainly resistance within Earthwatch to the partnership, as you know, because of the reputational issues of RIO TINTO and a number of colleagues at Earthwatch were members of Friends of the Earth and Greenpeace, where RIO TINTO has been regularly vilified rightly or wrongly for certain kind of things that it's done. And we had a series of presentations from RIO TINTO people to Earthwatch staff, which presented an opportunity to ask questions and to sort of … to challenge the company over certain issues. Maybe we didn't challenge enough … but it was a good process to go through. And it raised one question, I remember someone saying, "Why did you destroy so much the rain forests in Madagascar?" and the response being "Well we haven't! (Interviewee, Earthwatch)

According to the interviews, despite the fact that the debate within the organisation was a broadly democratic process and that some of Earthwatch's employees were members of radical ENGOs it is however questionable as to the extent to which they would have been able to adequately challenge the company due to Earthwatch's organisational identity. According to Dutton and Dukerich (1991:550)

---

[2] BIE stands for Business in the Environment.

"individuals have a stake at directing organisational action in ways that are consistent with what they believe is the essence of their organisation". Earthwatch employees, equally reflecting the organisation's identity and characteristics, would not be in a position to challenge Rio Tinto. The decision of both the company and Earthwatch to work together was based on the characteristics of Earthwatch, i.e. being a collaborative NPO.

The internal debate was indeed a difficult period for Earthwatch, testified to by the fact that all interviewees mentioned it as a significant event.

> The creation of this partnership when it became a sort of a bigger thing, we went through a lot of internal debate, we also brought RIO TINTO people down to present to the whole staff – because it was seen as a major thing for the whole staff and we, they needed to be convinced – and we didn't go through sort of a technical risk assessment and we certainly weren't and aren't in the position to sort of evaluate the specific environmental risks at places, as you know. But on a sort of generic level we had lot of debates, and a lot of debate with them present, where they would come to present what they were thinking and what they were trying to do and we encourage all our staff to fire some out questions and to have a proper good debate with these people. And ... because it was very important to us that right from the outset that was a sort of open road bus to discussion with the company so that they could see the kind of reservations that we as an organisation had, and that we could understand them and their responses. And I think, its not a technical risk assessment but it was a good process to go through, cause I think it also served to get the individuals on both sides who needed to work together, get them understanding better what the other's constituency was. (Interviewee, Earthwatch)

Even though the informal-internal risk assessment was an important process for Earthwatch it was the lack of a formal risk assessment process that potentially could pose risks to the organisation. It appears that Earthwatch did not consult with any radical NGOs, who were primarily the ones raising concerns about Rio Tinto's practices. In fact, an interviewee at Earthwatch accepted the fact that Earthwatch did not check with other NGOs outside their remit before entering the partnering process with Rio Tinto:

> I have to say our research wasn't very thorough.... I think the only NGOs we spoke directly to would have been other NGOs in sort of our end of the spectrum of the environmental movement and we would talk to them and said.... Like WWF but also, you know, people like the Natural History Museum and so on. And we would have asked them first whether they thought there were particular problems with RIO TINTO, whether they've done vetting themselves and so on. But secondly, whether if we entered a relationship with RIO TINTO that would make them regard us badly. (Interviewee, Earthwatch)

Hence, two different kinds of *internal risk assessment* processes were followed: (a) among the urbanization's employees; and (b) between the nonprofit organisation's employees and the employees from the company. Both processes seemingly appeared as an opportunity for employees to challenge openly the pre-partnership formation process and the organisation's decision to partner with Rio Tinto. Ultimately, however, it was the organisation's CEO that made the final decision. Hence, employees as internal stakeholders participated in the decision-making process mainly by not raising strong opposition; they accepted the senior management's decision.

> I can't say that it was a democratic decision, because in the end the chief executive needed to decide but on the other hand if there had been strong opposition from the employees, of

course, it would have been impossible for Earthwatch to enter that relationship. (Interviewee, Earthwatch)

The relationship was even further tested by members' who chose to resign from the organisation:

… we do have letters from members who resign, you know we've had people resign from us because of working with RIO TINTO.… (Interviewee, Earthwatch)

In 2003 Rio Tinto was still considered:

… as a controversial company among environmental NGOs and a number of Earthwatch staff and trustees raised concerns over the forthcoming partnership and its potential impact on Earthwatch's reputation. (Earthwatch paper, 2003)

Earthwatch's determination and strong belief in its organisational mission and the staff's identification with the organisation's mission made it possible to continue with the original decision, as an Earthwatch employee explains:

Earthwatch is always, we're always trying to talk about ourselves as being part of a wider movement, and because I think we're very conscious that we have one particular approach, but that particular approach, on its own, in isolation, would not be particularly effective. However, given the kind of spectrum of organisations that are out there, I think our approach is very important because I think if organisations lobby, if the company's lobbied hard and well it will decide to change, that's the objective of the lobbyists. But when it decides to change, the lobbyists aren't the right people to hold their hands and move them through the next steps; you need different kinds of organisations for different points along that road.… (Interviewee, Earthwatch)

The synergetic effects of the environmental movement are portrayed as significant factors in achieving cultural change within corporations. The multiplayer effect of different levels and directions of pressure is seen to provide significant leverage for potential change. According to Bendell's classification (2000c), Earthwatch belongs to the nonprofit organisations that promote change in corporate behaviour, through cultural and policy change. These changes, which will be discussed in detail in the next chapter where change is seen as a partnership outcome, refer mainly to the company, assuming a one-way influence and, hence, change. However, they do not pay enough attention to the fact that the two-way symmetrical communication may affect the nonprofit organisation as well, and even society as a whole. In the first case co-optation is a threat to the organisation. In the second case, society as an important constituent is being exposed to a multitude of messages, which can cause confusion. An Earthwatch executive questioned the clarity of the green message in an article that appeared in the "Third Sector" magazine:

There are undoubtedly a wide range of serious environmental issues that need to be addressed urgently, of which climate change is one of the most important. It is also true that there are different opinions with regard to how to address them, which I believe leads to the loss of a clear message to encourage positive and significant action. In this sense the green message can and does get lost. (Third Sector 2001:35)

The messages from the environmental movement are conflicting not only across the spectrum of organisations but also within the same organisation. As Crane

(2000:176) points out: "confronting the need to address cultural disharmony within, as much as across, collaborating organisations, is essential...".

Another type of risk assessment that Earthwatch performed was across other pro-business nonprofit organisations in the UK, such as WWF, and the Natural History Museum. In fact, Earthwatch, by performing an *external risk assessment*, relied: (a) on the risk assessment processes other organisations undertook before they engaged in a relationship with Rio Tinto; (b) on their answers to the question of how would they see Earthwatch if the organisation decided to partner with the company. The above demonstrates the *"partnership culture"* in Britain, as was the case with the corporate philanthropy culture in the US in the 1980s: "embedded in a network of relationships between corporations" (Himmelstein 1997:5). This culture is similarly shared among NPOs of similar convictions. Himmelstein, describing the corporate philanthropy situation in the US in 1997, argued that it is largely a political activity manifested within "a shared set of understanding about how philanthropy serves corporate interests, to whom donations should be made, and how giving programmes should be organised" (ibid). He concluded that it also manifests "a broad understanding of the place of large corporations in American society and a strategy for securing that place, which corporate philanthropy shares with other corporate political activities as well". Similarly, although now partnerships appear as apolitical processes in fact the selection of NPOs is a reflection of the dominant position of large corporations within the British society. Partnerships have replaced corporate philanthropy by scaling up the level of sophistication in the social involvement of corporations which is supported by a broad network of organisations and people that share the same social understanding of what is right and wrong. Remarks by an interviewee at Earthwatch appear in agreement with the above syllogism:

> And one of the things that encouraged us I think was that the Natural History Museum had a very big relationship with RIO TINTO ... in which RIO TINTO was sponsoring the Earth Galleries at the Natural History Museum, and I think we felt that if the trustees of the Natural History Museum and the Natural History Museum with all its resources had been through that sort of decision-making process, then it was that much safer for Earthwatch.

NPOs are determined to safeguard their survival and institutional position. In the case of collaborative NPOs, their organisational characteristics and mission statements do not place any mandate restrictions upon the partnership logic. Hence, when the organisation was offered the opportunity to partner with a large multinational corporation it would have been very difficult to reject such an offer:

> ... but, remember Earthwatch in those days was a very small, very financially insecure, very naïve organisation in many ways, on the other hand it was young and enthusiastic and could bring fresh ideas to, you know, the issues of partnership and so on. So, I think Earthwatch probably did things in its early days, like most organisations, that it wouldn't have done in its mature years, but there are benefits to that as well as, you know, downsides.

Figure 4.1 summarises the partnership selection phases that took place in the case of the Earthwatch-Rio Tinto partnership.

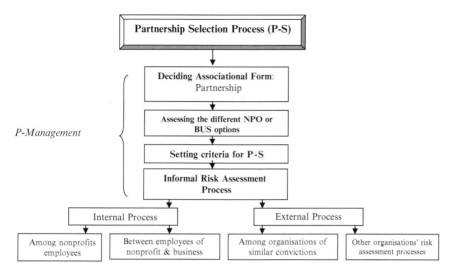

Fig. 4.1 Partnership selection: Earthwatch-Rio Tinto partnership

4.2.2 Partnership Design

The second phase of the relationship's evolution is the *partnership design*, which involves experimentation with the partnership relationship by setting objectives, drafting the Memorandum of Understanding (MoU), followed by a review process as suggested by an interviewee at Earthwatch.

Within the different phases of the partnership process the *management of the partnership* is a major issue that involves several processes such as partnership reporting and partnership structure, but also several departments in both organisations. In the case of Earthwatch, the Development Department was the one that originally developed the relationship into a partnership and the Corporate Programmes Department is the one that supports and expands all corporate programmes. In the case of Rio Tinto the External Affairs Department developed the relationship and it was also responsible for the implementation of the partnership with the parallel involvement of the Human Resources Department. The partnership has been repeatedly described as "resource intensive" as it required a diverse resource mobilisation, in two continents (due to the global character of the partnership), and involved four offices: Rio Tinto UK and Australia, and Earthwatch Europe and Australia. The content of the partnership is presented in Appendix 3, which indicates the spectrum of operations and resources that the partnership required. An important point that was raised by both partners is the resource imbalances that create expectation problems for the NPO. In fact, as pointed out in an Earthwatch paper: "Resources – the majority of NGOs are seriously smaller than their corporate partners, making it difficult to provide the sort of 'service' that a company might expect from the relationship" (Earthwatch 2003:2). The comment of a Rio Tinto interviewee on the same issue was: "I have my doubts whether a lot of NGOs can actually deliver ..." According to the interviewees the resource, expertise and professionalisation imbalances were a reality for the partnership and part of the complexity of cross-sector relationships. As the partnership management pertains throughout the different phases and stages of the partnership, each feature is being discussed briefly during each phase.

The partnership design is the partnership experimentation, which involved drafting the MoU and agreeing partnership objectives.

The first MoU was signed in July 1999 and took 24 months to develop. The MoU was signed by the Chief Executive at Rio Tinto Plc and the Chairman of Earthwatch Europe, manifesting the interest and leadership from the top of each organisation. A second and revised MoU was also signed when the partnership was renewed for a further 3 years. As RTZ was a global company, it was important that its partners were operating on a global level as well. The company tried to devise a unified MoU, but halfway through the process realised that due to the legal differences[3] in each country it would have been impossible. Hence there were two different MoUs, one signed in Australia among the partners and one in the UK. The above example demonstrates one of the important practical differences between sponsorship and partnership in the UK. As an interviewee of Rio Tinto mentioned the "customs and excise people have not yet caught up with partnership"; in effect partnerships are not yet being regulated in the UK. However, according to a Rio Tinto interviewee MoUs are indeed legally binding documents.

The Earthwatch-Rio Tinto MoU plays an important role in setting out the partnership objectives, which were discussed and agreed among the partners when the MoU was negotiated for the first time but equally when it was renewed in 2002. The objectives of the partnership are demonstrated within an Earthwatch Report to Rio Tinto (Earthwatch 2003R:1):

- Increase scientific data and improve conservation outcomes on a range of issues relevant to both Earthwatch and Rio Tinto globally
- Enhance scientific capacity in developing countries
- Increase awareness among Rio Tinto employees and other stakeholders of the role played by Earthwatch and Rio Tinto in promoting positive environmental outcomes
- Increase public understanding of biodiversity and the issues surrounding conservation

---

[3]One of the differences is that in the UK if you describe in the MoU the fact that one of the objectives of the partnership is for the company's benefit, then the partnership is considered a sponsorship and therefore is subject to VAT. This is not the case in Australia.

As demonstrated above there was no explicit articulated objective related to change in policy or practices of the company or indeed any objective mutually achieved by both organisations by combining their unique capabilities for the social good.

Partnerships are assumed as equal relationships between two partners, unlike sponsorships or donations. In those latter relationships reporting is usually one-way, i.e. the nonprofit organisation reports to the corporation in order to account for the money it has received. In the Earthwatch-Rio Tinto partnership reports took place only one-way, from Earthwatch to Rio Tinto. This can be interpreted as an asymmetry that demonstrated the transactional elements of the relationship demonstrated through the power dynamics that favoured the BUS partner. It also highlights the lack of a full appreciation of the concept of partnership.

The analysis of the reports further suggests a temporal language change. What is interesting is that in an earlier report to Rio Tinto, Earthwatch wrote: "However, the broader strategic aim of the partnership is to build the capacity of both partners to understand and manage biodiversity. Therefore the success of the partnership should be judged by whether this is being achieved" (Earthwatch 2001R: 3). And the report continued: "In addition, there are other aspects of the partners to be considered: Earthwatch wishes to have a positive impact on the way Rio Tinto operates regarding the environment in order to fulfil its own mission of environmental conservation" (ibid).

In later reports however this broader aim was not used in evaluating the partnership. And in the same paragraph the consideration on Rio Tinto's behalf was: "Rio Tinto needs to believe that Earthwatch is the right organisation to work with in terms of expertise, ability to deliver and being able to maintain a constructive relationship" (ibid). And although Earthwatch's employees were confident that in its "mature years" the organisation would have the confidence to address issues with more vigour, in fact 5 years into the relationship these statements were still not included in the report.

As these reports were exchanged only among the partners tentative suggestions as to why this happened could be: (1) that the two organisations reached the required level of trust and they did not feel the need for such statements due to the increase in familiarity and mutual understanding; (2) co-optation did take place and Earthwatch did not raise any further concerns about fulfilling its own mission.

Rio Tinto appears not to have an ad hoc approach in developing partnerships, instead its relationships are apparently well-researched, structured and carefully delivered and reviewed. An important aspect of the relationship is: "feeling comfortable with each other", as pointed out by an interviewee at Rio Tinto. However, at the same time perhaps the feeling of comfort can induce complacency. It seems that the internal risk barriers to the relationship were removed after the first 3 years of the partnership as trust increased.

Another similar point was raised again in the 2001 Earthwatch report to Rio Tinto:

> Earthwatch is a non-confrontational organisation which supports constructive engagement of the corporate sector. However, the organisation is unequivocal in aiming to have a positive influence on corporate behaviour and wishes to feel this is being achieved with Rio Tinto. (Earthwatch 2001R:8)

Again this consideration was removed from later reports. Also it is important that the same paragraph mentions a Rio Tinto consideration:

> As a global company Rio Tinto wishes to identify and work with NGO partners and needs to consider if Earthwatch is an organisation which it can work with in terms of expertise, ability to deliver and being able to maintain a constructive relationship.

As this report was produced towards the end of the first 3 years of partnership (2.5 years into the relationship), and nothing similar is mentioned in later reports, it can be assumed that in order for the partnership to continue to exist Earthwatch was successful in maintaining its 'non-confrontational' character by sustaining a largely constructive relationship with Rio Tinto. However, contrary to the above, Earthwatch officially had the right within the partnership's MoU to exercise criticism to Rio Tinto:

> ... They are essentially in the businesses of the protection of the environment. I wouldn't even suggest that any of these relationships are designed to stop them from criticizing us. In fact, there's a clause in the agreement that specifically preserves the right of either party to comment on the other. In other words, there is a clause that says that nothing in this agreement shall be taken as limiting the rights of either party to comment publicly on the other. Because a lot of our relationships are with environmental NGOs who do speak publicly and critically sometimes. (Interviewee, Rio Tinto)

Despite the above clause in the MoU in reality it appeared that only in extreme circumstances the NPO partner would choose to do so. In other words it would be very difficult for an NPO to criticise the company while in partnership. Also the motives of the organisations manifested through the partnership's objectives in the reports turned from being vocal and explicit to silent. In the earlier reports each partner feels it important to demonstrate what each needs; however, what is really missing within the reports is the 'we' mentality that is a significant characteristic of the partnership philosophy. This will be further discussed in the discussion section, but also within the final chapter of the book.

Access was not always granted in order to pursue further research. Equally, due to the complexity of this subject it is only possible to give indications based on the material to hand rather than offer definite answers to questions. Based on the above, the extent of the criticism Earthwatch directed at Rio Tinto during meetings is not known. However, Earthwatch's interest to continue the partnership for as long as possible was manifested by the fact that the above report[4] recommended that the partnership should be renewed for 5 more years.

Another interesting aspect of the partnership is that although the partnership did include some planned publicity the company "has been quite reluctant to publicise too much" as this could draw more attention to the relationship. A press release was jointly drafted and sent out to the charity and industry press, avoiding the broad sheets; equally some advertisements about the partnership featured in specialised media.

---

[4] A report that was authored by Earthwatch exclusively for Rio Tinto.

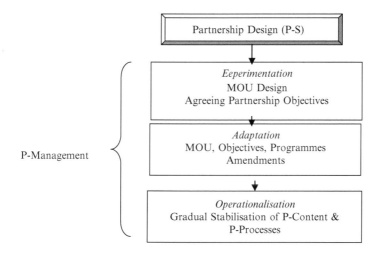

Fig. 4.2 Partnership design stages: Earthwatch-Rio Tinto partnership

The next stage of the partnership design is the operationalisation of the partnership, which flags the level of agreement that has been reached between the two partners with regard to the programmes and processes involved. The partnership management encompassed all three stages (experimentation, adaptation and operationalisation) and also included the programme management of the activities required to support the work of the departments involved in the implementation. Moreover it included the partnership review meetings, which took place twice a year in order to monitor and evaluate the progress of the partnership.

The second partnership phase is presented in Fig. 4.2 which summarises the partnership design.

4.2.3 Partnership Institutionalisation

At the time the data was collected, partnership institutionalisation was the next challenge for the Earthwatch-Rio Tinto partnership. Although the relationship has been tested through crises and survived, the organisational actors expressed reservations when talking about the institutionalisation of the relationship. They appeared to believe that the relationship is more dependent on personalities than on the level of institutionalisation.

> Ideally, one would like these things to be institutionalised so that the personalities don't matter, but I think, inevitably, they do. There has been a change of person in Earthwatch and partnership is no longer what it was. It's no longer as dynamic and creative. So, yes, it does matter. (Interviewee, Rio Tinto)

However, even prior to the change in personnel this level of institutionalisation was probably difficult to achieve as the two organisations did not appear to work

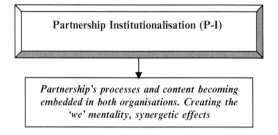

Fig. 4.3 Partnership institutionalisation: Earthwatch-Rio Tinto partnership

together on a project of mutual interest that would probably assist in the level of institutionalisation and the development of the 'we' mentality which is a feature of partnership relationships (Austin 2000) (Fig. 4.3).

Finally, one of the programmes that is integral to the partnership is the 'The Global Employee Fellowship Programme' (EFP), through which Rio Tinto employees worldwide volunteer to participate in global conservation projects funded by Earthwatch (for a description of the programmes see Appendix 3). The EFP has an impressive personal impact on the company's employees[5] ranging from increasing their interest in the environment and conservation issues to submitting their resignation to the company upon their return. However, this impact is limited to the numbers of participants. Out of a total of 36,000 Rio Tinto employees worldwide, the total number of fellows (140) that were placed on the EFP programme represents 0.4% over a period of 6 years, a percentage that could not be responsible for a cultural shift by itself. The comments on the reports are indicative of the cultural shift that the EFP initiates and that potentially could lead to organisational change through people. In fact, in cases where the number of employees volunteering would increase then the partnership has the potential to gradually change employees' perceptions if indeed the charities address the issues of change strategically.

4.3 Partnership Phases in the Prince's Trust-Royal Bank of Scotland Case Study

The second part of the chapter presents the interactions within the second case study. The different phases that were identified in The Prince's Trust-Royal Bank of Scotland Group partnership are: partnership selection, partnership design, partnership institutionalisation, and change through partnership as presented below.

---

[5] Each employee completes a questionnaire on returning from their placements. These comments are compiled annually by Earthwatch and are included in the report to Rio Tinto.

4.3.1 Partnership Selection

Partnership selection is the first stage of the relationship which started when the Royal Bank of Scotland Group made the decision to develop partnerships rather than other associational forms in order to achieve the company's strategic objectives. It was a conscious decision that aimed at developing partnerships across different organisations. Unlike the approach of most companies as funders of the nonprofit sector, the RBSG was interested in establishing a relationship that would allow for an evaluation of the bank's contribution to its partner. Hence, the bank's question to The Trust: "where would you like to be in 3 years?" Presumably The Trust offered a satisfactory answer to this question, which allowed for the actual partnership process to commence. The bank went through a partnership selection process in order to identify potential partners.

> ... obviously over the last three years of being first of all, defining what it is we want to do, identifying the right partners and then putting in place a series of projects.... (Interviewee, RBSG)

However, the relationship with PT was not new for either of these recently-merged banks (RBSG and Natwest), consequently the partnership selection process was rather brief.

The deliberation of the partnership meant that it was part of a planned rather than an emergent process and it was based on a number of criteria. The level of familiarity between the two organisations was an important factor as explained in the quote above. According to PT interviewees among the reasons that RBSG chose the nonprofit organisation were: (a) both organisations had strong royal affiliations; (b) both organisations operated in the same geographical areas within the UK and both had headquarters in London and Scotland; (c) social exclusion was a priority for both organisations; (d) PT was a professional NPO; (e) it provided a safe profiling platform for the company; (f) both organisations had an interest in business start-ups; and (g) the bank offered personal development opportunities through the PT volunteering scheme which resulted in setting the same objective to reach a target of 550 volunteers from the RBSG. On the other hand, the PT decision was also based on the fact that the commonalities across the two organisations and the opportunity to receive substantial funding would allow for changes to take place, which was one of the motives for the partnership relationship.

Hence, arriving at the partnership decision did not appear to be a difficult task between the two partners. There was no evidence of risk assessment process prior to the partnership, similar to that which Earthwatch underwent in the partnership selection stage. PT is a NPO that did not pose any foreseeable risk for the bank. RBSG did not go through any risk assessment process either:

> No. We didn't on our side, no. I mean because we know, you know, your gut feels is probably as good as anything in terms of is this right and we knew the partnership had already been set up, so it seemed we had to just go with the flow.... (Interviewee, RBSG)

There is no evidence that PT went through a formal or informal risk assessment process prior to the decision to form the partnership. However, the interviewees

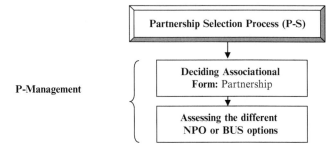

Fig. 4.4 Partnership selection: PT-RBSG partnership

suggested that PT has a risk assessment process in place as described in the quote below:

> ... we chose our partners carefully and we've got a risk assessment process ... so we've got that process in place.... And one of those is that we do not accept funding from partners that don't fit our objectives. (Interviewee, PT)

Figure 4.4 summarises the partnership selection that took place in the case of the PT-Royal Bank of Scotland partnership. The boxes below describe the processes.

4.3.2   Partnership Design

The second phase of the relationship's evolution is the *partnership design*. This stage involved experimentation within the partnership relationship by agreeing the partnership objectives and also through the exchange of a letter called a 'heads agreement':

> We have a letter.... It sets out the terms under which we're handing over the money. What the money is for and it goes to Route 14–25 or ... but it's not a legal contract in a sense.... You could probably call it heads of agreement.... (Interviewee, RBSG)

Although staff in both organisations were clear about the total amount of money and the different facets of the partnership, it seems that there was confusion around the issue of the existence or otherwise of a contract. Some appeared to think that there was not a contract at all and that the relationship was 'personal and informal' and others believed that there was a contract among the partners. Although this issue is not considered of major importance for the implementation of the partnership, it seems that PT arrived at the realisation that it could perhaps offer an answer to the following observation:

> So we probably should have involved more people, more operational employees at an earlier stage I think,.... (Interviewee, PT)

An important element of the partnership design is the establishment of the 'virtual team', as it is called across both organisations. The employees of PT and RBSG comprise the virtual team and act as counterparts in each other's organisations. The 'virtual team' allows for: (1) multiple points of reference within each partner's organisation consisting of a widespread network of people; (2) development of trust among the virtual team members and in effect among the two organisations; (3) avoiding over-centralised power on one or two individuals in each organisation; (4) better operationalisation of the partnership. Another important element of the virtual team the background of its members.

> ... both (name of RBSG employee) and (name of RBSG employee) come from ... backgrounds which have involved community work..., so they're not pure bank backgrounds.... Actually that's been extremely helpful to me and I think to other people in the organisation.... (Interviewee, PT)

As it becomes apparent the virtual team consists of a 'new breed of executives', an interesting addition to the way partnerships are being operationalised and ultimately assist in improving the understanding across sectors but at the same time increasing institutional isomorphism. Figure 4.5 demonstrates the relationships across the two virtual teams. For example, circle number six in PT represents the 'Communications manager' that has a direct relationship with circle six which is the 'Media relations manager' at RBSG's virtual team. The reason they share the same colour is because they mirror each other's responsibilities with regard to the partnership and in fact my impression from the interviews was that the RBSG counterparts acted almost as second line managers to the PT counterparts. The numbers represent the range of different functions within the partnership team of each organisation.

The shades behind each number represent the effect that the individual can have in its own organisation but also due to the interaction with each other's organisation. The shades in the case of RBSG are upright representing the focused effect an individual's role can have within a profit sector organisation of the structure and size of RBSG. On the other hand, the shades are sideways in the case of PT indicating the more general effect that the role of an individual in a nonprofit organisation of

Royal Bank of Scotland Team   The Prince's Trust Team

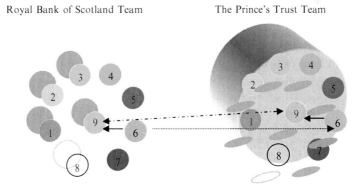

Fig. 4.5 The virtual teams: PT-RBSG partnership

a smaller size (in comparison with RBSG) can have. Hence, it is faster to deliver, but also to observe change in a relatively small organisation as the effect of the individual's contribution becomes more apparent. The change is faster and more generalised. In the case of a big for-profit sector organisation although the change in BUS takes place it needs more time to become manifest. The change is delivered through the individual's effect on the organisation; hence it needs larger participation of members of staff to deliver a generalised change. Also the generalised change of the nonprofit sector organisations relates to their proximity to the social arena due to their mission. Their deliverables are directed to the social sphere hence they have the potential to influence society much more rapidly. Therefore, the interactions among the sectors are important but the nonprofit sector holds a higher level of responsibility towards society.

Number 9 in both organisations represents the role of Senior Project Manager at PT and the Community Investment Manager at RBSG. Both are central to the partnership function as they play a central role in the partnership. These two people are informed about everything (with regard to the partnership) within their own organisation (e.g. line 6–9) but also in each other's organisations (e.g. line 9–9). The virtual team's role is also to introduce and understand each other's organisations, which is very important especially for complicated organisations. Hence the virtual team also functions as a steering wheel for enhancing the organisations' knowledge and understanding.

> I mean by big thing is I think we all do respect each other a tremendous amount. And sort of trust each other to do things. So ... so like people, micromanaging, you don't get that so much because we're all left to do our own set of things. And if I need advice on something or how to do something, I'll go to RBSG employee name and ask "what's the best way to do this within your organisation?" you know.... And talk things through. We do have a very open relationship with them. ... when things are sort of difficult ... and we work around them. (Interviewee, PT)

The virtual teams in each organisation represent an sub-culture which aims to integrate both its own culture in the other's organisation but also the other's culture in its own organisation. This could potentially cause tensions primarily within each organisation's own culture, which, in my opinion, depends on the degree the values that each team represents are embedded within the rest of the organisation. An example that demonstrates how close the members of the virtual team are to their partner's employees (chiefly PT's virtual team members) is that PT's Employee Investment Manager is called "PT in residence" by the RBSG staff. The level of commitment and integration of the particular individual resulted in achieving the partnership's target of 550 volunteers from the bank one year in advance of the original deadline. In addition to the virtual team there are three individuals within RBSG that have/had dual responsibilities across the two organisations. The dual responsibilities of are an indication of the multiple roles organisational actors play across different sectors. Hence gradually the different cultures or different perspectives that the two sectors represented are going to diminish due to the increased interactions across the sectors on a personal and organisational level. The relationship between the two partners may appear too close in the eyes either of their own employees or of other stakeholders, which might suggest that the distinctiveness of each organisation should be a priority, especially for PT.

Within the relationship the management of the partnership is a major issue that involves several processes such as partnership reporting and partnership structure, but also several departments in both organisations. In the case of the PT-RBSG partnership both organisations were actively involved in developing the partnership. However, some employees at PT perceive the partnership dynamics to be led by RBSG:

> ... in general I think it's on the borderline between them telling us what they think we should be doing and us driving where we think we should be going.... To be fair it's the same with quite a lot of our fundraising in that we invent things that aren't necessarily that we have thought about ourselves that were in our ... what we felt we needed to do but we dressed things up and package things so that they become more attractive in order to get funds and then we end up delivering something that was not necessarily in our original remit to the way we thought we were going to go. Now in this instance its actually turned out that where we've gone has been for the benefit to the organisation with the route 14–25, but some of the add-ons that we have laid on top probably have required more resource than the return we would have got. (Interviewee, PT)

The above remarks highlight the negative possible effects of a centralised team where the individuals at large agree with how the relationship is being managed, although ignoring the subculture that perhaps exists within the organisation as well as divergent opinions. Hence, although there were no indications of internal opposition at PT as was the case with Earthwatch, the existence of a subculture was evident which, if it had not been ignored, might have added an important dimension to the relationship.

Similarly to the Earthwatch-Rio Tinto partnership reporting practices, in the PT-RBSG partnership reporting existed also only one way, from PT to RBSG. The lack of a full appreciation of the form of a partnership relationship contributes to the continuation of a transactional relationship. Some of the RBSG interviewees appeared to expect their PT counterparts (even if both people were members of the virtual teams) to report to them as they would expect a paid agency to act. Also, although this issue was discussed in the past by one of PT's employees, it seems that RBSG did not take any further corrective action.

> I remember (name of PT employee) actually asking that once and I remember mentioning it to (name of RBSG employee) and nothing ever came of it.... I think it's because they're all doing that work on our behalf, or on the PT's behalf because it's helping the PT ultimately obviously.... It's kind of like a management report, which I think is fairly standard, you know, like when you're being given money you have to report back on how it's being spent, for example if there was a PR agency that I was giving loads of money to then they would give me a monthly report, ... (Interviewee, RBSG)

The process of the partnership did allow for several opportunities in order for issues to be discussed e.g. during the annual partnership review and the bi-weekly management team meeting across the two virtual teams; however, they did not allow for such issues to be resolved.

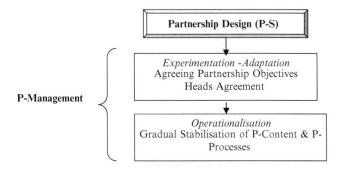

Fig. 4.6 Partnership design: PT-RBSG partnership

The content of the partnership is presented in Appendix 4, which indicates the spectrum of operations and resources that the partnership required. The partnership processes and content were discussed and agreed in the early days of the partnership. Hence, the relationship became fully operationalised at a very early stage. A contributing factor was the fact that trust already existed in addition to the background of the members of the virtual team. Hence, the partnership processes involved were brief (Fig. 4.6).

The next stage of the partnership design is the operationalisation of the partnership, which flags the level of understanding that has been developed between the two partners and the agreement of the programmes and processes involved. Figure 4.6 summarises the partnership design and its respective stages. We can observe in the partnership design stages of the PT-RBSG partnership that the first (experimentation) and second (adaptation) phases are incorporated into one phase (unlike in the previous case study). The reason for this is the strong organisational compatibility across the two organisations which allowed for a faster evolution of the relationship.

4.3.3 Partnership Institutionalisation

The next stage in the partnership evolution is the partnership institutionalisation, in which the partnership content and partnership processes become embedded in both organisations. When a relationship has been institutionalised it means that although crises occur, they can be resolved rather than cause a serious problem in the relationship. A number of interviewees referred to a total of 15 incidents that describe situations that could be categorised as 'crises situations'. However, both organisations dealt with these incidents in a 'mature' way that offers testimony to the quality, but also the stage of the relationship. Unfortunately, most incidents are classified as 'off the record' and thus quotes cannot be used for illustrative purposes. The quote below testifies to: (a) the fact that such incidents occur; and (b) the mature reaction of the bank:

> I think, you know, because they are very understanding, and they do appreciate that these things happen.... (Interviewee, PT)

This also testifies to the respect of each partner and its own incubation times until results are achieved or processes are put in place.

The 'we' mentality is central to the reactions of the organisations and the level of personal familiarisation that occurs in this stage of the relationship:

> ... these guys all work as one team and I went to a meeting and I couldn't have told you who was Royal Bank and who was PT, ... (Interviewee, RBSG)

Another observation regarding the central teams or virtual teams of the partnership is that they function as a 'breakwater' for the incidents that occur. Since the members in both teams have developed a higher level of familiarisation for each others' organisations their level of tolerance is higher among employees within the rest of the organisation. However, it appears that the same level of understanding is not shared across everyone in each organisation, providing further evidence for the existence of a subculture:

> It is very good and is very promising that in terms of their internal communications is not necessarily that ... not everyone shares that view, you know, (not) everyone is as progressive as the central team.... (Interviewee, PT)

The 'core people' of the partnership enjoy the familiarization benefits among which are that they: (1) ask each other for advice; (2) encourage new ideas and assist in their implementation; (3) increase their contacts by the available network in the partner's organisation. The familiarisation process is encouraged by both organisations actively. More specifically an 'away day' was organised that allowed the members of the central team to further develop their understanding and get to know other members of the virtual team rather than only their counterparts:

> And that was really nice and it was really good to see that ... is like really committed, really into it, really enthusiastic about what they are doing ... and we were able to sort of have a good laugh and we joked each other about stuff. And that was really nice.... And that was just for one day. And everyone I'm sure.... I think everyone came away from that feeling about although everyone was quite motivated beforehand I think they came away from that feeling.... Because they knew everyone who's kind of behind it.... (Interviewee, PT)

An important observation is that the partnership institutionalisation makes sense *on the personal level*, i.e. when the individuals develop a personal relationship of trust within the partnership then the level of embeddedness of the relationship becomes more evident. However, the partnership institutionalisation does not refer only to the agreement between the two partners, but also to their ability to disagree without causing termination of the relationship. The following quote offers an example of such an incident.

> ... and also there's one project where the bank had signed up, or part of the bank had signed up to do something with The Trust.... The Trust appointed the individual, but then had them do a different job and we said, well hang on that's not what we funded, so The Trust actually returned the money.... (Interviewee, RBSG)

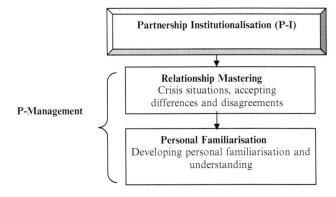

Fig. 4.7 Partnership institutionalisation: PT-RBSG partnership

The above incident testifies to the level and quality of the institutionalisation. Even when the organisations do not agree in their approach to a specific issue they are allowed to disagree and function in an autonomous way without consequences that might jeopardise the viability of the relationship. This is an important element of the partnership relationship: to disagree within the partnership.

Figure 4.7 summarises the institutionalisation stage of the partnership relationship. The first stage appears to involve overcoming crises in two ways: (a) by accepting the other organisation's strengths but also weaknesses as a reality of an integrative relationship; and (b) by not avoiding conflict but rather accepting disagreements as functional which permits retaining the organisation's identity intact. The first stage is termed 'relationship mastering' and the second 'personal familiarisation' involving the familiarisation on the personal rather than the organisational level. Examples include: incidents of spending informal time with each other were reported by interviewees such as having lunch together or staying at someone else's house, and so forth.

4.3.4 Change Within the Partnership Process

Within this partnership case study the level of institutionalisation is comparatively high compared to the previous case study. The relationship is far more embedded within both organisations which is evident by the use of the word 'we' and the partnership language employed by some of the members of the virtual team. Although key people in the partnership did not change within the first 3 years it seems that the relationship will be able to tolerate changes of staff due to the level of commitment from the highest ranks of the organisation and the durability of the relationship which was tested and prevailed through a series of events.

The PT-RBSG partnership developed with a clear mandate: RBSG's funding would allow The Trust to experiment with each work approach (both in content and processes) hence achieving organisational change through re-identification. The term re-identification is used to describe the process that PT undertook that resulted in changes in process and outcomes. This process comprises of a series of events and organisational processes that allowed for the re-identification to take place. Re-identification refers to the re-branding exercise that PT went through but also to the changes in its programmes as a result of the experimentation. Although I attempt below to summarise the processes and events, it is by no means a conclusive undertaking. As an organisational actor at PT also remarked about change: "these questions need to be asked in a year's time when changes could better be observed and understood". The interviews probably took place too soon after the events that were occurring in order to allow for a period of reflection for the organisational actors.

Change management at RBSG underpins many of its functions and programmes; change is embedded within the organisational characteristics of the bank since it has a long standing tradition in implementing change. This contributed to the strategic way in which RBSG facilitated change within PT. Furthermore a set of change drivers contributed to the rapid changes within PT: (1) the familiarity that existed among the two organisations due to the compatibility of the partners and the historic relationship that preceded the partnership relationship; (2) the compatible backgrounds of the people who were working for the virtual teams in both organisations; (3) the dual roles and responsibilities of 3 people in the higher ranks of the organisations.

The changes that took place and were identified at PT were: (1) changing the way of working; (2) changing the target group of The Trust; (3) changing the core programme and as a result developing the programme 'Route 14–25'; (4) initiating a re-branding exercise; (5) the appointment of new people within The Trust. The above issues manifest the multidimensional character of change. The first dimension of change was on the personal level, which then acted as leverage for the organisational, identity and process changes. Also, the general change was the product of: (a) the process change that was a result of the partnership evolution process and could be characterised as unintentional change; and (b) the output change which was a primary objective for RBSG and also for PT, hence it can be termed 'intentional change'. The intentional dimension of change is demonstrated by the quotes below:

> We nested a series of, I suppose, second order objectives, the primary objective was we believed that our cash injection would help the PT achieve a shift in the way it did things. (Interviewee, RBS)

> We do need to change, we desperately need to change, because ... we, it's a competitive market, you know, and every market's competitive, we have to be more effective at reaching our target groups. That is the number primary objective.... (Interviewee, PT)

As this case study exhibits, corporations can also facilitate change within NPOs and hence Bendell's (2000c) conceptualisation with regard to NPOs facilitating change within BUS needs to be extended in order to include the facilitation of change from BUS to NPOs. On the other hand, the intentional changes took place also within the commercial partner: changes were mainly at the human resources (HR) level. The Education and Professional Development Manager developed an HR matrix that matches the personal development programme of RBSG's staff to the volunteering opportunities of PT. The changes that accrue from the interactions between PT and RBSG are on a personal level and hence can be characterised as subtle and silent and were primarily unintentional:

> Maybe we have changed … maybe it has changed, but I mean I think it would be subtle and over a period of time.… (Interviewee, RBSG)

The change that takes place at the bank is a process that equally is/will be driven by people. The total of 560 volunteers that RBSG offered to PT is a very small percentage within the Group. It represents 0.63% of the total RBSG UK workforce. On the other hand, RBSG's volunteers represent 80% of PT's workforce and 5.6% of its volunteers. However, while the numbers of volunteers will increase, and hence the time that the volunteers are exposed to the PT culture, the partnership is expected to deliver a more generalised attitude change, similarly to the one that the Earthwatch employees were expecting from the volunteering programmes.

Comparing the change between RBSG and PT we can observe that: (1) change is symmetrical, i.e. it takes place in both organisations; (2) the quality of the change is different hence the changes at PT can be characterised as intense and concentrated in a short period of time; on the other hand the changes at RBSG are more subtle but are expected to have temporal impacts within the bank. The different types of changes are presented graphically in Fig. 4.5 (The Virtual Teams) in the section P-Design through the shading of the circles.

The complete model of implementation is presented in Fig. 4.8. The four phases within implementation (selection, design, institutionalisation, change) are followed by exit strategy; although exit was not evident in either of the case studies under investigation it would be the final implementation phase in case of the relationship's termination. The partnership review process can take place as part of the exit strategy or during the institutionalisation phase and can feedback and inform all phases of the relationship (Seitanidi and Crane 2009).

The section discussed the change processes that are summarised in (Fig. 4.9). The two types of changes that occurred – processual change and outcome change – indicate that change can be examined both as a process but also as an outcome. The above section discussed change as part of the implementation stage hence as a process.

Within the next chapter change is discussed as a partnership outcome.

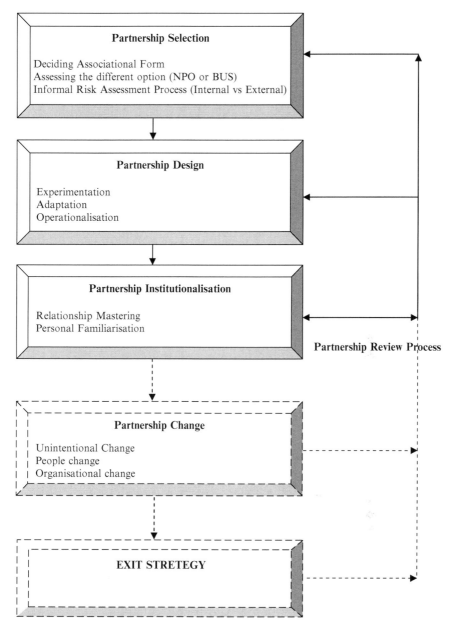

Fig. 4.8 Partnership phases

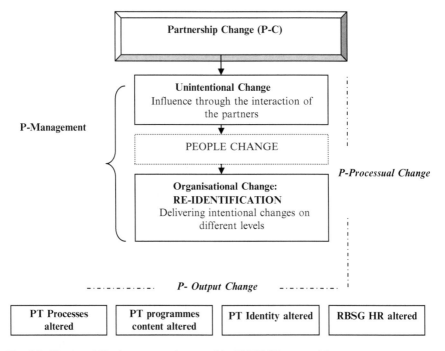

Fig. 4.9 Change within the process of partnership: PT-RBSG partnership

4.4 Conflict in Captivity

This section discusses the above findings in light of the 35 comparative interviews that were conducted within 29 organisations in order to confirm, disconfirm or extend the findings of the implementation stage in partnerships.

Within the partnership selection phase both case studies suggest that the partnership was initiated by the BUS partner based on the level of familiarity that stemmed from the organisational characteristics that pre-existed between the organisations and a number of commonalities (e.g. operating in the same geographical areas, compatible organisational objectives). The BUS partners did not feel that it was necessary to undertake any risk assessment before the implementation of the partnership, unlike the NPOs. This testifies that, unlike NPOs, BUSs do not feel any threat or the anticipation of conflict when considering to partner with a collaborative NPO. Earthwatch undertook a formal and informal risk assessment process. However, based on the interviews conducted for this research, it appears that the majority of NPOs and BUSs operate on a 'gut feeling' when they choose their partners. This indicates a strong informal side of conducting the selection process that relies on perceptions and a predominantly subjective appreciation of situations rooted within a *network culture* of organisations with similar perceptions about their role and relations to other sectors.

Based on the above there are two parameters that relate to the collaborative NPOs and the existence or absence of conflict. The first parameter is that collaborative NPOs do not pose any inherent threat for their BUS partners that can lead to conflict. The second parameter is the extent to which any form of threat remains captured internally within the partnership relationship rather than being externalised within the social arena. There is only partial evidence suggesting that NPOs collaborate with across the spectrum of the nonprofit sector, i.e. environmental NPOs working with non-environmental NPOs (based on the comparative interviews e.g. Greenpeace working with Oxfam) or collaborative NPOs requesting advice from confrontational NPOs, hence exchanging information within the social arena. In the case of Earthwatch the organisation only contacted environmental NPOs that were also collaborative, i.e. within the same network sharing similar perceptions. Hence the criticisms towards Rio Tinto were not addressed but rather subjectively assessed by a small number of individuals who were also not experts in the areas required to make informed assessments (such as in mining as suggested by Partizans). However the risk assessment process and the existence of a sub-culture within NPOs reveals a *covert side of conflict* within the relationship. The partnership design and institutionalisation phases provide evidence suggesting the existence of a *sub-culture that remains captured and embedded in compliance* with the dominant organisational culture. The sub-culture that exists within organisations was demonstrated in an individualistic rather than a collective mode hence it remains as an isolated expression of opinion.

During the implementation of the partnership the conflict remains captured through the dynamics of the two partners and the asymmetrical resources. In the case of collaborative NPOs they do not appear able to manage the asymmetries as suggested previously by a number of authors (Uri 1991; Covey and Brown 2001; Hamman and Acutt 2003) either by developing strategies or communicating warnings to their BUS partners (ibid). The reasons appear to be the *lack of previous experience in managing confrontational approaches, the financial basis of asymmetries resulting in the dependence of NPOs and the internalisation of any form of conflict.* Hence the partnership might result in a 'lock-in effect' and remove from them the possibility of criticism in a silent way (Shaw 1993, cited in Millar et al. 2004:406).

Consequently although there was a clause in the MoU of the Earthwatch-Rio Tinto partnership allowing Earthwatch to criticise Rio Tinto it would have been impossible for Earthwatch to publicly criticise their partner because of the consequences they would be faced with: (a) risk losing the funding from Rio Tinto or further extension of the partnership; (b) risk losing the financial support of other BUSs that funded Earthwatch; (c) having to shift Earthwatch's mandate and hence the organisation's identity. The above was further confirmed by NPOs far more experienced in confrontational approaches, such as Greenpeace:

> And when you work in collaboration with people you have to be prepared to compromise and work towards the common goal.... The tough thing is ... getting what you want without compromising your core values and beliefs.... We still absolutely negotiate everything! ... and sometimes it's tough, you know and we all get tired. (Interviewee, Greenpeace)

One of the *silent risks* of partnerships can appear from within the partnership, namely 'negotiation fatigue' (as the quote suggests above) which can in fact pose challenges to both organisations. Even organisations with strong political positions such as Greenpeace can experience the pressure for conformity within the partnership relationship.

However, the partnership between WWF and Lafarge appeared to disconfirm the above and present a different picture of the potential that a collaborative NPO might hold. WWF moved beyond its collaborative approach to use confrontation and *publicly* challenge a corporation that works in partnership with, reclaiming its NGO identity and responsibility. In fact, in 2003 WWF UK returned its share of a £3.5 million funding to Lafarge, a multinational aggregates company "because they refused to abandon plans to build the UK's biggest superquarry on an unspoiled Scottish island" (Third Sector 2003). The partnership was between WWF International and Lafarge; WWF UK opposed the quarry before the beginning of the partnership, therefore they rejected the money as they did not want to create confusion about their position. The website of WWF Scotland remarks in 2004:

> Since the start of the international partnership between WWF and Lafarge in 2000, WWF has consistently and fully supported the actions taken by Scottish NGOs opposed to the quarry (LINK Quarry Group), and has repeatedly affirmed its strong opposition to the proposed 600 hectares quarry.... The history of this controversy demonstrates that engaging in partnerships with companies does not prevent WWF from criticising and opposing any controversial aspect of our partners' activities. WWF believes that it is important to engage with business and industry in the push towards a more sustainable future, and will continue to seek out partnerships that, it strongly believes, can contribute globally to significant social and environmental benefits. (WWF 2004)

In April 2004, WWF's website welcomed the decision of Lafarge not to pursue its interests further in the development of the quarry. The above is one incident where WWF, a collaborative NPO, *publicly challenged a partner and returned money, hence placing the conflict in the social arena.* So far it appears to be the only UK charity that used this approach publicly, although Greenpeace UK made similar remarks about their intention to pursue a similar tactic:

> ... companies need to know that if they don't move far enough that people are going to come back and they're going to campaign. (Interviewee, Greenpeace)

The change in WWF's political position marks an important shift in one of the traditionally collaborative NPOs as it emphasises the need for a 'weapon' that collaborative NPOs need to employ in order to challenge corporations and push them towards change:

> You don't negotiate unless you have some weapon that you can use. In those negotiations you don't have a power to negotiate. What in fact happened in many instances was that yes from time to time the corporation discussed issues with the NGO. But they began to set limitations in the way the NGO could operate. If the NGO criticised then the company professed to be angry, because we are engaged into discussion with you. So, that process of co-option in fact silenced quite a lot of NGOs. (Interviewee, International Textile, Garment and Leather Workers' Federation)

The ability of an NPO to *externalise the conflict* by increasing the reputational risk for the business partner appears to be important in order to facilitate *organisational change* in BUSs. However, it seems possible only when (a) there is no financial dependency from the BUS partner; (b) the NPO has a strong positive reputation and (c) is well-established, as was the case with WWF. This was further confirmed by the incident that occurred within the second case study when PT returned money to RBSG due to a disagreement on the role of a funded position by RBSG.

The implementation stage revealed the *different expectations* of the BUS partners with regard to the capabilities of their nonprofit partners to meet the implementation demands. Hence both Rio Tinto and RBSG provided *funding for positions* crucial to the implementation of the partnership objectives. In effect this can increase the dependency, the expectations and the workload for the NPO. Furthermore the dynamics between the partners with regard to the reporting practices and meetings and the overall process that was followed appeared to be led by the BUS partners. It can be suggested that BUSs have more experience in partnerships hence they take the lead in similar situations. Also the financial dependency of NPOs might create barriers in explicitly taking the lead or arguing for different processes.

The *partnership institutionalisation* is a phase that permits the partners to master the relationship both on organisational and personal levels. Accepting each partner's strengths and weaknesses allows for a *maturation phase* succeeded by accepting *conflict as functional* within the relationship. Within the first case study the existence of opposition to the partnership relationship revealed conflict but this was removed through the formal and informal risk assessment process that took place as part of the partnership selection stage. There were no further incidents that offer evidence of any form of conflict. Within the second case study there was an incident that allowed PT to assume its role and responsibility in order to pursue its own mission as it saw fit. There were incidents in both cases, as presented within the chapter, that revealed how the partners accepted the strengths and weaknesses of one another, however with regard to conflict it appears that in the majority of cases it is seen as dysfunctional and therefore avoided. One possible explanation is the strong monetary element in partnerships and the instrumentality of achieving the desired organisational outcomes by both partners.

Finally, there are two types of change that take place within the partnership implementation: *intentional* and *unintentional*. Evidence of the intention for change appeared originally through the motives; as an idealistic motive in Earthwatch and as an instrumental motive for both PT and RBSG. Within the second case study (RBSG facilitating change in PT) the intention for change was carried through a strategy during the implementation. In the first case study there was no indication of a strategy within the implementation stage. The unintentional change that takes place refers to the processual change which is part of the partnership process and the people change which occurs through the interactions among the organisational actors. The facilitation of change is linked with the experience of the partners in similar processes and their motivation for the partnership relationship. In the case of PT it appears that the organisation did not include in its motives the facilitation of change within RBSG. However, one of the main interfaces among the partners was to assist disadvantaged young people, the main target group of PT, which are excluded by high street banks receiving loans. Since PT had the experience, it was in a position to facilitate intentional change within the policies of the banks.

The comparative interviews further clarified the link between the lack of conflict within partnerships and change as suggested by the following quote:

> I think one of the interesting things was that there was not enough conflict in the relationship.... Because we were all too nice to each other.... (Interviewee, Deutsche Bank UK)

If indeed conflict remains captured within the relationship then the potential for change that stems from the different perspective the partner brings into the relationship is minimised. As an interviewee of the International Textile, Garment and Leather Workers' Federation (representing 10 million workers in 130 countries) pointed out, there is a need for different perceptions that can induce change in companies both in the UK and the US:

> Here, only a couple of weeks ago in a leading corporation the guy responsible for CSR said: 'the only time I can see my CEO is when we are threatened with an exposure in televisions, or newspapers, or something'.... That highlights to me and I heard it repeated in the US how NGOs must continue to be advocates, to be campaigners. Because most of these corporations will only change under pressure. And they have to be brought to change, they will be brought to change kicking and screaming and opposing it.... The real benefits come from the NGO exerting continuously pressure, highlighting where there is wrong and forcing change into the company. In that way the NGO maintains its integrity and at the same time it is achieving results. The company itself is also benefiting, you know, not getting into that Margaret Thatcher situation. You know, everybody around the table agrees with you all the time. You know the famous story when she was 10 years in power the cabinet had a dinner and the waiter said 'what will you have to eat?' and she said 'I am having the steak' and he said 'and vegetables?' and she looked around the table and she said 'they will have the same as I do'. (Interviewee, International Textile, Garment and Leather Workers' Federation)

4.5   Conclusion

The implementation stage of partnerships involves the following phases: selection, design and institutionalisation. If intentional change is carried through a strategy facilitated by either the profit or the nonprofit partner, change can follow as the next phase in the relationship. If there is no strategy in place then change is primarily unintentional and takes place in a subtle way, unlike the intentional change that is intense and concentrated within a short period of time.

It appears that due to the importance of the informal side of the relationship and the financial dependency of the collaborative NPOs, any form of conflict remains captured within the partnership. However, the case of WWF demonstrates the potential of externalising the conflict and in effect forcing BUS to change. The interactions among the partners during the implementation stage are led primarily by the BUS partner. The instances when the NPO play a more *assertive role* explicitly demonstrate that there is the potential for a more critical and active role that collaborative NPOs should assume as it has the potential to balance the power asymmetries among the partners.

Chapter 5
Stage Three: Partnership Outcomes

5.1 Introduction

The third empirical chapter examines the outcomes that accrue as a result of the partnership relationship. It examines each case study separately by looking into the organisational outcomes for each participating organisation. The research concentrates on positive outcomes of partnerships. The organisational outcomes are firstly discussed within each case studied followed by a discussion on the difference between social and societal outcomes. The types of change as a partnership outcome are presented within each case study and later discussed within the comparative interviews that were conducted for the research across 29 organisations.

5.2 Organisational Outcomes in the Earthwatch-Rio Tinto Partnership

This section explores the positive outcomes that accrued from the first partnership case study for both Earthwatch and Rio Tinto. The categorisation and analysis of outcomes presented below is based on the organisational actors' statements during the interviews.

Both organisations recognised that the partnership outcomes met their expectations and in some cases even exceeded them by presenting unanticipated outcomes as a result of the process. These are termed 'processual' outcomes as they derive from the process of the partnership relationship. On the other hand, the 'content' outcomes refer to the anticipated outcomes that were part of the motivation for the initiation of the partnership relationship.

Tables 5.1 and 5.2 summarise the outcomes for Earthwatch and Rio Tinto respectively as identified by the interviewees.

The processual outcomes occur due to the relationship as the familiarity and proximity of the partners increases and allows for more opportunities for beneficial

M.M. Seitanidi, *The Politics of Partnerships: A Critical Examination*
of Nonprofit-Business Partnerships, DOI 10.1007/978-90-481-8547-4_5,
© Springer Science + Business Media B.V. 2010

Table 5.1 Earthwatch's partnership organisational outcomes

	Content outcomes	Processual outcomes
Tangible outcomes	Financial support	Additional financial support
		Technical expertise on IT and health and safety
		Develop new partnerships
		Capacity improvement
Intangible outcomes	Profiling Earthwatch	Learning experience and evolution of the organisation
	Support for organisation's mission (research and education)	
	Use the company's influence as leverage to exert influence within the business sector: stamp of approval from Rio Tinto	Increased familiarity and working relations with Earthwatch Australia
		Exert more political power within the Earthwatch network of organisations
		Networking opportunities

Table 5.2 Rio Tinto's partnership organisational outcomes

	Content outcomes	Processual outcomes
Tangible outcomes	Employee engagement	Developing a constituency within the environmental sector
	Access to environmental experts	
	Shape environmental and biodiversity policies	
Intangible outcomes	Demonstrate commitment to employees for environment, shared values of corporation and employees	
	Increased employee morale	
	Profiling Rio Tinto as a leader within extractive industries: stamp of approval from Earthwatch	Trust from NGOs
		Better understanding of the NGO sector
		Build up a degree of good will in quite important players in environmental sector (buffer zone)

exchanges. For example, shaping the company's policy on biodiversity was never included in the partnership objectives; however, according to some of the interviewees' perceptions shaping the company's policy on biodiversity did take place due to the existence of the partnership relationship. An Earthwatch employee testifies to this in the following quote:

> Sure, there is just one thing that I would add that I think, something that wasn't in the partnership very strongly originally but it's become ... became more prominent was the issue of Rio Tinto needing advice specifically on the area of biodiversity.... That was probably in the period about 1999 I would think, perhaps 1999–2000. (Interviewee, Earthwatch)

The second dimension of the outcomes refers to the extent to which the outcomes are of *tangible* or *intangible* nature. The intangible outcomes refer, for example, to the image of the partner, the networking opportunities, the improved understanding of either the profit or the nonprofit sector. An example of an intangible outcome for Earthwatch and in particular of exerting more political power within the Earthwatch network of organisations is demonstrated by the following quote:

> … this wasn't at all Rio Tinto's objectives – but it brought Earthwatch Europe and Earthwatch Australia together much more closely, bound by this partnership and it meant that our strength and our motivation in sort of increasing our influence in the Earthwatch family was vastly greater. So, what you see now, is an …, in the Earthwatch family you see a very powerful Earthwatch Australia and Earthwatch Europe, and a less powerful Earthwatch US and that in part was the result of the fact that Rio Tinto brought us together.… (Interviewee, Earthwatch)

As shown by the above quote the intangible outcome can also be classified as processual since it was not originally included within the content of the partnership relationship. Hence, we observe a relationship between the perceived motives of the organisation and the outcomes. The perceived anticipated outcomes are the attained motives as they consist of the content outcomes. The unanticipated outcomes are additionally achieved outcomes that derive from the process of the relationship. In the case of the Earthwatch-Rio Tinto partnership there are no unattained motives apart from the idealistic motives expressed by Earthwatch employees. Although they were articulated as motives they have not been achieved hence they consist of unarticulated outcomes as demonstrated by the quote below:

> … we maybe don't capture the benefits of those but they carry on. It's a great thing, you can't capture it, but just because you can't capture it, it doesn't mean that is not to be encouraged! (Interviewee, Earthwatch)

Earthwatch's unarticulated outcomes "make a difference within the business community", "change-influence companies by policy advising" refer to the social and societal outcomes, which appear as more difficult to be captured and strategically enforced by the organisation. They will be further discussed in the section on social and societal outcomes below.

The main tangible and content outcome for Earthwatch was the funding provided by Rio Tinto. The tangible processual outcomes were: technical expertise on IT and Health & Safety issues that was transferred by Rio Tinto; additional financial support received through the years; development of know-how on partnerships that allowed Earthwatch to develop similar relationships; capacity improvement that was developed with Earthwatch due to position Rio Tinto funded. The intangible and content outcomes encompass: improving the organisation's image, in particular within the corporate sector, and receiving indirect support for the organisation's mission as a result of the relationship.

On the other hand, Rio Tinto's tangible and content outcomes include HR opportunities for its employees with positive results such as enhancement of employee morale (intangible), improved access to environmental experts and assistance in shaping the environmental and biodiversity policies. The interviewees referred to only one tangible processual benefit which was developing a constituency with the environmental sector through Earthwatch's relationships. The intangible content outcomes included the demonstration to employees that the BUS is committed to the environment and hence the individual values of the staff are shared by the company (which was one of the main concerns of Rio Tinto as described in the previous chapter) hence resulting in increased employee morale and profiling Rio Tinto as a leader within the extractive industry with regard to its environmental position. According to the interviewees this was achieved through the association with Earthwatch. Finally, the intangible processual outcomes included increased trust within the NGO community but also improved understanding of the NGO sector for Rio Tinto, hence an improved level of good will towards the company within the environmental sector.

Another observation is that according to the perceptions of the interviewees there is a clear concentration for Earthwatch on processual outcomes, unlike Rio Tinto where there is a concentration on content outcomes. This might offer a suggestion that the positive outcomes or benefits for the NPO derive predominantly from the process, the major tangible one being the financial support they receive from the BUS. For Rio Tinto less intangible benefits appear based on the perceptions of the interviewees which might offer an indication of the real tangible value of the relationship for the BUS which lies within the HR function and the benefits that derive from the employee engagement.

The above findings will be discussed further within the discussion section of the chapter also in light of the comparative interviews as some of the outcomes refer to assumptions within parties external to the relationship dyad (i.e. the environmental sector).

The next section will discuss the outcomes for The Prince's Trust-Royal Bank of Scotland case study.

5.3   Organisational Outcomes in the Prince's Trust-Royal Bank of Scotland Partnership

This section explores the positive outcomes that accrued from the second partnership case study for both The Prince's Trust and RBSG. The outcomes that are categorised below and analysed by the researcher are based on the organisational actor's statements during the interviews.

Both organisations recognised that the partnership outcomes met their expectations and in some cases even exceeded them by presenting unanticipated outcomes as a result of the process. The classification of outcomes and explanations provided above in the Earthwatch-Rio Tinto case study is applied below. Tables 5.3 and 5.4 summarise the outcomes for the organisations as identified by the respondents.

The intangible outcomes, which are different for each partner, are of enormous importance to both partners as testified by the quotes below:

> So, those sorts of intangible things are obviously, have immense benefit to us, and it's amazing how you kind of learn, you know, and benefit from that, ... so, those sort of intangible

things, you know, are hugely beneficial and sometimes you don't even realise that you've got them until they're not there.... (Interviewee, PT)

It's so difficult to compare, we give them all this money and what they give us back is kind of intangible, but yes it helps us in enormous ways on a number of different levels, right down from the staff.... (Interviewee, RBSG)

The intangible and content outcomes for PT refer to image (profiling PT), learning through the partnership relationship, and gaining support for the organisational mission, exerting influence within the business sector and hence improving understanding of BUSs, getting more networking opportunities, strengthening the 'power' of the PT regions through the association with local RBSG and Natwest branches and getting closer to the private sector's way of thinking. The intangible and content outcomes for RBSG refer similarly to: enhancing the image and brand name of the bank as a responsible company, and having long-term social impact within society. On the other hand, the tangible and content outcomes for RBSG encompass HR development opportunities which is one of the most substantial outcomes, along with the accrued benefits related to staff retention, recruitment and morale development; develop more contacts with high net worth individuals that are part of the PT funding and relationship basis, meeting the government's agenda and hence improving their access to government

Table 5.3 The Prince's trust partnership organisational outcomes

	Content benefits	Processual benefits
Tangible outcomes	Financial support (£3.7 million) and nature of funding (allowing experimentation and change); long-term support	Additional financial support (£60,000)
		Consolidation of all loans
		Develop new partnerships
		Increased resource to the PT regions (regional offices)
		Technical expertise on IT and personal tracking system within organisations, financial discipline (performance frameworks)
		Capacity improvement
Intangible outcomes	Use the company's influence as leverage to exert influence within the business sector: stamp of approval from RBSG	Increased power to the PT regions
		Learning experience, evolution, opportunity to innovate
	Support for organisations' mission (volunteer and social inclusion)	Networking opportunities
	Profiling PT in the media and enhancing its brand name (increased)	Private sector way of thinking about management
	Change of the organisation	
	Improved understanding about corporations, different culture exposure	

Table 5.4 RBSG's partnership organisational outcomes

	Content benefits	Processual benefits
Tangible outcomes	HR development opportunity in a unique unconventional way through PT's programmes (Matrix development matching core competencies development with the volunteering opportunities at PT)	Learned how to operate within cost-risk lending and potentially develop a new market Loan business New Bank accounts
	Improve staff development, moral, retention and recruitment	Assist in bringing together Natwest and RBSG
	Develop more contacts with high net worth individuals	
	Second generation customers (expanding the bank's customer base)	Developed a model that can be transposed onto different organisations
	Meeting the government's agenda and achieving better access to government	
Intangible outcomes	Profiling RBSG in the media, softening the image of the bank and enhancing its brand name (good strong recognition at a local level; increased CSR benefits): stamp of approval from PT	Improve the community investment return ratio (qualitative measures)
	To have long-term social impact (create a legacy), which will result in a better off society hence a sustainable business environment	Better understanding of the NGO sector, different culture exposure Networking opportunities

(PT has a strong institutional relationship with government) and finally developing second generation customers through the young people that PT supports. However, according to an RBSG interviewee the commercial benefits come as a second priority to the bank. It is interesting that although the loan facility to PT was given with no preferential terms RBSG's interviewees did not mention it as a positive outcome for the bank.

The tangible outcome for PT with regard to the content of the partnership is the financial support of £3.7 million it received. With regard to the processual and tangible outcomes the Trust received additional financial support; a consolidation of all its loans was achieved (due to the existence of the relationship); the partnership know-how allowed it to develop new partnerships; the resources of the regional offices of PT were enhanced due to the institutional relationship (through the local RBSG and Natwest branches); IT expertise was transferred from the bank to the Trust resulting in a number of positive developments (e.g. personal tracking system for PT 'customers'); and finally capacity improvement through the funded positions by RBSG.

An example of the intangible and content outcomes for RBSG is the networking opportunities within a particular target group for the bank: the government sector.

An example of how the relationship with PT facilitates the networking opportunities for the bank is demonstrated by the following quote:

> They get the opportunity to meet other organisations who may be, you know, that we can introduce them through the awards and the events that we actually have. I mean we will have a number of sort of potential, maybe some people from the government, so for example last year at the national event there were some quite key figures from DWP, you know Department for Work and Pensions and some quite key figures from the government whom we have been trying to get closer to, came to the event. Now RBS obviously because they were at the event, then they can get ... we can introduce them.... (Executive, PT)

The link between the motives and the outcomes can also be observed as indicated in the previous case study. The anticipated outcomes are the attained motives and the unanticipated outcomes are additionally achieved outcomes. In the case of the PT-RBSG partnership the intrinsic and instrumental motives were attained according to the perceptions of the interviewees and in fact exceed their expectations. The idealistic motives that were present in the RBSG's interviewees – 'do the right thing' and achieve 'social benefits' – were attained through the support they offered to the Trust. The section on social and societal outcomes will further discuss the above.

It appears that based on the perceptions of the interviewees PT attained more tangible processual and content intangible outcomes. This can be an indication that not only the content of the partnership was beneficial but also the process of the relationship. The lack of more tangible content outcomes might be an indication of the lack of previous experience that would allow for more content outcomes to be expected and act as motivation for the partnership relationship. For RBSG all different categories of outcomes seem more well-balanced testifying to the experience of the bank with similar relationships and also the pro-active approach they employ.

The above findings will be discussed further within the discussion section of the chapter also in light of the comparative interviews as some of the outcomes refer to assumptions within parties external to the relationship dyad. The next section discusses the social and societal outcomes of both partnerships.

5.4 Social and Societal Outcomes in NPO-BUS Partnerships

This section moves beyond the organisational outcomes that accrue through the partnership to examine the social and societal outcomes of the two case studies.

The focus of each partnership predominantly reflects the NPO's area of expertise. In the first partnership the focus was biodiversity and conservation and in the second partnership social exclusion, in particular, of disadvantaged young people. The aim here is not to analyse the programmes of each organisation, which would be beyond the scope of the book, but rather to use the perceptions of the interviewees on positive outcomes (benefits) that accrued from the relationship in order to explore the social and societal dimension of the outcomes.

As neither partnership employed indicators in order to assess the outcomes, and in particular social indicators, the motivation for the relationship is used in order to examine the original intent for delivering outcomes for society. The *social outcomes* predominantly accrue from the role of the NPO. In the case of Earthwatch the social outcomes were: being able to continue their work on biodiversity through the financial support of Rio Tinto (in all countries the organisation was involved, such as the African fellowship programme that Rio Tinto supported); the improvement of Earthwatch's Health and Safety operations resulting in higher safety measures for all participants of Earthwatch's projects; and finally delivering more informed and sensitised corporate employees with regard to environmental issues as a result of their participation in Earthwatch's projects.

In the Earthwatch-Rio Tinto partnership, as pointed out earlier, the idealistic motives indicated that the organisation wanted to make a difference within the BUS community by influencing corporations and their policies. This would entail a *societal outcome* "shaping environmental and biodiversity policies" which would benefit society as a whole and in particular the environment. However, there are some important points to consider: (a) the partnership was not developed around this theme as it was pointed out that this issue emerged through the process – even when the advisory role interface appeared, Earthwatch did not demonstrate any intent accompanied by a strategy that would facilitate such a process nor was the organisation able to assess the outcomes; (b) Earthwatch employees did not mention it as a positive outcome of the partnership; (c) some of the Rio Tinto interviewees mentioned it is as a positive outcome of partnerships; however; (d) Rio Tinto interviewees did not appear to attribute to Earthwatch such an outcome, even if it was achieved by the company:

> Well some of it may have a direct link, I mean it may be involved less so with Earthwatch, if we take FFI or Bird Life, it might be directly related to an area where we're operating Kew Gardens is another one. Well there is work going on directly related to an area where we're operating, it can help us with rehabilitation techniques or an understanding of the biodiversity of the area in some ... probably less the case with Earthwatch at some of those. (Interviewee, Rio Tinto)

What in fact was attributed to Earthwatch was:

> But there are intangible benefits for us, in a sense that working with organisations such as Earthwatch gives an understanding to people within the environmental biodiversity community of a mining company on what we do.... You see there is something within an organisation that they're actually nervous about working for us. I mean they have a perception of a mining company, an extractive industry company in a particular way.... Once they get to know it through the partnership, then they have a much greater understanding of what we do and see the practical conservation environmental protection activities that we engage and see our sense of environmental responsibility. And they then share that with others, and they then talk about it....And sharing experiences of what RIO TINTO is doing. Equally using ... we're using their expertise to develop our thinking and share our ... and as we develop for example a biodiversity strategy in that we can call on a whole series of people outside of the group to get their comments, their opinions, their input, and I think that helps us. (Interviewee, Rio Tinto)

Furthermore, there seemed to be no project that was developed as a result of the partnership relationship that was different to Earthwatch's existing programmes or that solved a problem that could be considered 'in part as social' (Waddock 1988:18) and that "extend(ed) beyond the organisational boundaries and traditional goals and lie within the traditional realm of social policy" (ibid). Hence, it appears that the relationship between Earthwatch and Rio Tinto fostered a 'partnership approach' with strong transactional elements (one-way reporting, financial support to existing programmes) but could not be classified as a social partnership. The relationship had clearly social outcomes due to the work of Earthwatch but did not go beyond the remit of either organisation combining their unique capabilities in order to offer a solution to a social problem. Furthermore the lack of *intent, strategy* and *assessment* for the societal outcomes could potentially weaken the importance of the relationship for the organisations but also for society and reserve the outcomes as only *anecdotal feedback*:

> So, we would have made an impact on that level but, a lot of that kind of stuff is anecdotal feedback rather than a properly evaluated studies of the impact we're having because we can't just do that. We don't have the resources and also we wouldn't be able to do it. (Interviewee, Earthwatch)

Within the second partnership the social outcomes that accrued from PT's expertise were: the ability not only to continue their work on social exclusion for disadvantaged young people but also to experiment and ultimately to 'change' due to the financial support provided by RBSG; the improvement of their technical expertise on IT and financial discipline that resulted in improved services to the target groups the PT serves; and finally increasing the awareness of RBSG's employees with regard to social exclusion but also offering more specialised advice to young people through the volunteering opportunities within the programmes of the Trust.

In the PT-RBSG partnership the idealistic motives appeared within the interviews of the bank, advising 'achieving social benefits' or 'doing the right thing'. Both of these refer to social benefits that accrue from the support they offered to the Trust. These have been achieved as remarked above. The indication of societal outcomes derived from the CSR dimension that pro-actively is practised by the bank:

> And ultimately as an organisation we're successful because the communities in which we operate are successful. So, it's also we're part of that community, we're not a separate, stand-alone kind of ... we're actually in the heart of those communities that we're supporting and it's important that they do well and as a result we do well.... I think society part, the society benefits, partner benefits in terms of The Prince's Trust and business benefits for the bank.... (Interviewee, RBSG)

Although there is a clear awareness of the need to support the communities in which the bank operates, and a strategy that successfully delivers support to the NPOs, it lacks the *reflective approach* in how to address problematic social issues that encompass both the bank's operations (and in effect the banking sector) and the Trust's capabilities through the partnership relationship. Neither the bank nor the Trust addressed, for example, the issues of adapting the bank's policy in offering loans directly to disadvantaged young people. This entailed an excellent opportunity for both organisations to address this issue through the partnership and create a blueprint for other banks to follow. Until the time the data was collected for this case study there was no indication that they started working towards this end. Similarly, there were no other projects developed together by both organisations that extended beyond the organisational boundaries and their traditional goals or indeed that lie within the traditional realm of public policy (Waddock 1988). As was the case in the previous partnership examined, the PT-RBSG relationship has fostered a 'partnership approach' with transactional elements but could not be classified as a social partnership, according to the definition provided by Waddock (1988).

This section discussed the social and societal outcomes of the partnership relationships making a distinction between the social and societal outcomes. Based on the types of outcomes that were achieved, based on the perceptions of the interviewees, it was suggested that the partnerships that were examined could not be qualified as social partnerships. The next section discusses organisational change as a partnership outcome.

5.5   Change as a Partnership Outcome

In the previous chapter change was examined as part of the partnership process. Within this section change is seen as a partnership outcome. As remarked by a number of authors change is conceptualised as an interaction field (Abbott 1992; Lovelace et al. 2001; Heracleous and Barrett 2001) and the aim is to look at a holistic explanation of change. Hence the previous chapter unavoidably mentioned the types of changes that took place, in particular within the second case study, as it was a natural process that followed after the partnership institutionalisation. This allowed for contextualising the process of change (Pettigrew 1985) and placing it within the historical context of the relationship. Within the first case study change was mentioned within the institutionalisation process.

In this section the discussion concentrates on the *content of change* or in other words 'what has changed' on the organisational level. There were two types of change[1] that took place. Firstly, the *deep level change* that leads to changes affecting the 'core' of the organisation (Barnet and Carroll 1995:223) which will be referred to here as *organisational genesis*. Genesis refers to the change in substance according to Aristotle (cited in Peters 1967:116–117) and which he defined as the "passage to the enantion"[2] (ibid: 71). Secondly, the opposite would entail a significant difference which would be against the pre-existing condition or the predominant 'status quo' (Martin 2000). Organisational genesis in partnerships would entail different choices that lead to different actions (Martin 2000) that affect the core of the organisation.

---

[1] These types of change do not refer to the intention for change.

[2] *Enantion* means 'opposite' in Ancient Greek. In Heraclitus "there is an essential unity ... of the opposites, a unity that it is not obvious, but that maintains the unity-plurality in the opposites" (Peters 1967:52).

Hence, a substantial change in the core of the organisation to take place requires two different entities to partner in order for them to influence the other. The second type of change refers to 'non-core changes' that occur in the periphery structure Barnett and Carroll (1995:224). These changes are termed *organisational kinesis*. According to Aristotle (cited in Peters 1967:116–117) when the changes take place in "the three categories of quality, quantity, or place, the *metabole*[3] is called *kinesis*". Organisational kinesis or non-core changes can take place in the form of process or outcome changes.

Within the Earthwatch-Rio Tinto case study organisational genesis did not take place due to the focus of the partnership on the organisational level and as the interviewees did not report any core changes in either organisation. If the interviewees had attributed to Earthwatch the change of Rio Tinto's biodiversity policy and the implementation of the policy was perceived as an outcome, then it would be considered as organisational genesis. On the personal level however the Earthwatch 'EFP' (Employee Fellowship Programme) of the partnership programme was reported to have had a profound impact on a number of Rio Tinto employees who resigned after returning from their participation in Earthwatch's programmes. This was an *episodic type of change* (Weick and Quinn 1999) that indirectly affected the organisation. As explained in chapter four the total number of employees that participated in the EFP represented only 0.4% of the Rio Tinto employees[4] hence the overall impact was minimal.

On the other hand, the non-core changes, or organisational kinesis, that took place referred to: (a) a new position that was funded within Earthwatch by Rio Tinto; (b) the improvement of know-how on IT and Health & Safety issues at Earthwatch; (c) increased contacts from both organisations within each other's networks; (d) potentially altered perceptions about both organisations within a number of target groups; (e) better working relations between and across the organisations involved in the partnership (Earthwatch UK-Rio Tinto UK; Earthwatch UK-Earthwatch AU; Rio Tinto-UK-Rio Tinto AU). Overall the above changes were unintentional and primarily can be classified as processual changes that occurred in the course of the relationship.

Within the second case study, change was intentional and it was RBSG that facilitated the process within the PT. The change drivers for the second case study were presented in Chapter 4. The changes that took place as a result of the partnership and which were identified by PT's employees, consisted of: (a) the changed way of working; (b) the changed target group of the Trust; (c) changing the core programme and as a result developing the programme "Route 14–25"; (d) initiating a re-branding exercise that led to the re-identification; (e) the appointment of new people within the Trust. These five changes represent a profound change within PT, according to the interviewees; hence they consist of organisational genesis as the initiated changes resulted in the introduction of a new programme, target group and new identity of the organisation that produced

---

[3]Metabole means 'change' in Ancient Greek.

[4]The number of employees who resigned is not known, but during the interviews the use of the plural indicated several people.

different results. These changes were *intentional* as the motivation for change existed in both organisations hence they consist of content change further classified as core changes consisting of organisational genesis. The above do not present an assessment of the changes, but rather are based on the perceptions of the interviewees. The *first dimension of change* was on the *personal level*, which then acted as leverage for the *organisational* (identity and process) *change*, indicating the multidimensional character of change.

The people change, however, was not reported to be a profound type of change as was the case with Rio Tinto's employees. Also the processual change that took place was unintentional, unlike the output change (the five above reported changes) which were intentional. The intentionality of change is demonstrated by the quotes below:

> We nested a series of, I suppose, second order objectives, the primary objective was we believed that our cash injection would help the PT achieve a shift in the way it did things. (Interviewee, RBS)

> We do need to change, we desperately need to change, because ... we, it's a competitive market, you know, and every market's competitive, we have to be more effective at reaching our target groups. That is the number primary objective.... (Interviewee, PT)

Furthermore, organisational kinesis took place by: (a) the improvement of IT know-how and other internal systems of the Trust; (b) increased contacts for both organisations within each other's networks; (c) new HR opportunities within the bank; (d) potentially altered perceptions about both organisations within a number of target groups; (e) better working relations between and across the organisations involved in the partnership (PT Headquarters – RBSG Headquarters; regional PT offices with regional branches of RBS and Natwest;). Figure 5.1 summarises the types of changes, giving at the same time examples from both case studies.

Although working in a partnership relationship for many years might allow each organisation to get closer and become familiar with each other, assessing change in each other's organisation is not an easy process:

> I think, well, yes, Rio Tinto appears to have changed, appears to be a constant changing culture, Rio Tinto appears to be taking these things more and more seriously, and I think what is hard in a partnership with our resources and the way we structure our partnership, is to answer that comment there, at site level. (Interviewee, Earthwatch)

Also the unintentional change that takes place is not easily assessed as it occurs continuously with long-term impacts, hence it might remain unobserved for a long period of time:

> ... actually is a really good place to work (PT) and it helps their retention rates, it helps them feel part of their community and they're not just doing their job to make other people rich, the rich get richer, they're actually affecting change within their local communities.... (Interviewee, PT)

The last section of this chapter discussed change as an outcome within both partnerships; the next section will discuss the outcomes of the partnerships within the comparative interviews.

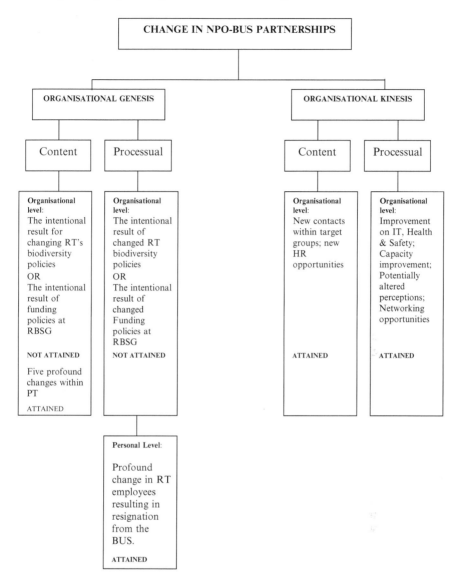

Fig. 5.1 Change in NPO-BUS partnerships

5.6 Change in Captivity: Convergence of Need Versus Divergence of Missions

In both the cases examined the organisational outcomes present the most prominent type of outcomes. This testifies that positive outcomes can accrue as a result of the partnership for both organisations on the organisational level, hence the "mutuality of benefits" (Wilson and Charlton 1993) is evident. The organisational outcomes that took place within the second case study appear to be greater compared to the first. This can be indicative of *the level of institutionalisation* of the partnership relationship in particular through the processual outcomes in each case. There is a higher concentration of outcomes for the BUS partners, in particular within content outcomes which could be an indication of the level of experience in organisations in forging relationships.

With regard to the *content of the organisational outcomes* for the profit sector the most valued outcome appears to be the HR opportunities, increasing the reputation and credibility of the organisation and access to target groups and hence improving their image; also to develop second generation customers for RBSG and receive advice for the formation of biodiversity policy in the case of Rio Tinto. On the other hand, the most valued organisational outcomes for the NPOs were receiving financial support, receiving technical expertise on a number of issues (IT, Health & Safety), and increasing access and credibility within the business community, hence enhancing the NPO's name recognition. For both sectors the outcomes do not appear to differ in content from the outcomes reported by previous studies (Berger et al. 2004; Austin 2000; Heap 1998; Greenall and Rovere 1999).

It is important to note that all the above organisational outcomes represent the perceptions of the interviewees and are not verified or audited outcomes. There was no evidence in any of the above case studies that the organisations developed any partnership indicators (Caplan and Jones 2002) against which the outcomes could be assessed. As such this is an underdeveloped area both in theory and practise.

The comparative interviews emphasised the significance of the organisational outcomes for both organisations. The importance of the following tangible/content organisational outcome for NPOs was underlined: the financial support received from the BUS. An employee of the RBSG repeated what was reported as a question that many NPOs asked their BUS partners: "tell us what you want us to do and we will do it". This stresses the importance of financial resources for the nonprofit sector and offers an indication of the willingness to collaborate with BUS with unknown levels of resistance. On the other hand, the most valued organisational outcome for BUSs appears to be the *HR opportunities through volunteering* within the nonprofit partner as it addresses the most important target group of BUSs i.e. the employees. As reported in the 'Third Sector' magazine the above is confirmed by a survey study: "This suggests that charities value income above all else, and that these partnerships are a good way for companies to muzzle charities', said Joe Saxton, co-founder of nfpSynergy, the think-tank that conducted the survey" (Third Sector 2005a). As a result the societal side of partnership becomes less prominent and as a WWF interviewee remarked: "most partnerships are opportunistic".

The partnership case study between WWF and Aviva, a non-financial partnership that did not involve the exchange of any monetary resources, disconfirmed the above as it aimed at an informal exploration into the understanding of sustainability between individuals across the two organisations within the financial sector in order to improve their practises. Although this partnership did not have a high level of institutionalisation, *achieving intentional outcomes was extended beyond the organisational level to the sectorial level*, which could potentially lead to *societal positive outcomes.*

In both cases, according to the perceptions of the interviewees, there is a higher concentration on processual outcomes, which was also confirmed by the comparative interviews both within BUSs and NPOs. The differences of tangible-intangible benefits were not confirmed or disconfirmed by the comparative interviews as the time constraints of the study did not allow for a further combination of analytical depth and breadth within the comparative interviews.

The segregation of the nonprofit sector creates gaps in perceptions that if addressed might allow better integration of the sector and closer collaboration among the different sections of the sector, and across sectors. One of these gaps relates to the perception of outcomes. For example, the Rio Tinto employees suggested during the interviews that one of the outcomes of the partnership with Earthwatch was "building a level of goodwill in quite important players of the environmental sector". One of the important players of the environmental sector is WWF. However, the perception of Rio Tinto's and Earthwatch's employees that important NPOs might change their minds about the company does not seem to be the case of informed 'players' within the environmental sector. When the WWF interviewee was asked if the historically collaborative NPO would form a partnership with Rio Tinto in the UK, his answer was an emphatic 'No', due to reasons explained below:

> Because they're one of the world's largest coal producers, I would not take money from RIO TINTO, punt, full stop, unthinkable, in my book.

Furthermore, when I suggested to the interviewee that Rio Tinto presents itself as a company that embraces change, the response was:

> That's outrageous! They're the world's largest coal producer, the largest ... huge contributor to CO_2 emissions, it's all green wash. We shouldn't buy it!

The interviewee did not comment about the partnership between Earthwatch and Rio Tinto, but reiterated the earlier statement.

Based on the above it appears far more difficult to alter the perceptions of organisations than suggested by the interviewees. The reason that the perceptions of Earthwatch and Rio Tinto interviewees appear diametrically opposed to the perception of the WWF interviewee is possibly an indication of the *lack of communication across organisations during all partnership stages.* This is the case even within the same spectrum (both WWF and Earthwatch are collaborative NPOs and both also have together a partnership with HSBC), which emphasises the *lack of transparency and openness.* The employees and their perceptions appear trapped within the culture of the organisations they are working for, in most cases, serving the interests of the prevailing organisational status quo. The aim is not to assess if indeed Rio Tinto has changed or not, but rather what are the perceptions of different organisational actors within and outside the partnership relationship.

All the organisations that participated in the research reported organisational and social partnership outcomes. The social outcomes derive from the role and mandate of the NPO (as similarly reported by Austin) hence *the claim of BUSs achieving social outcomes by association to NPOs.* These social outcomes are indeed beneficial for society, the NPOs and their work; however, they do not represent a unique contribution of both the BUS and NPO in solving a common social problem together. Both BUS and NPOs suggested during the interviews that they have many partnerships; however, when asked they could not refer to any common projects on which they work together with their partners, or report any social or societal outcomes. Hence it appears that the majority of organisations employ a *partnership approach* rather than build a *social partnership as a distinctive form of association*. The *partnership approach* can deliver more intangible and processual benefits due to the proximity and familiarity between the partners, unlike other approaches (transactional, philanthropic). The *partnership form* on the other hand requires partners to be working together on a programme or policy that addresses a social issue where the outcomes are not only benefiting the partners but are externalised to society i.e. delivering societal outcomes.

For the BUS, organisational outcomes entail the main reason to develop a partnership relationship with an NPO; hence increasingly what were previously termed 'social partnerships' are referred to as 'strategic partnerships' as they aim to satisfy the organisational needs through a partnership relationship. In social partnerships the primary aim is to *serve society* and not the organisational needs. In fact the social character of partnerships might indeed appear against the organisational needs as was the case with Lafarge and WWF. WWF UK, by returning the money to Lafarge, forced change within the BUS and in effect Lafarge changed their actions. However, it appears that in most cases *the politics of NPO-BUS partnerships fail the societal dimension of partnerships by capturing the potential of organisational genesis* (or core changes) *within the convergence of need rather than divergence of missions*.

The two different types of change – organisational genesis and kinesis – were confirmed by the comparative interviews. The decision of Lafarge, for example, to abandon the plans to build a super-quarry in Scotland represents a core change that was the result of a productive public disagreement between WWF UK and the company. The remarks below from a WWF's interviewee offer an insight on the core changes that HSBC had to undergo due to the partnership:

> Oh, achieving change. My starting point is to achieve change and at the same time we get in money ... that binds us together! The money, it's not because we can spend it on some fresh water projects, but it's more commitment on their side. You see HSBC giving us money is actually for fresh water requires from their perspective that they change their fresh water lending policies. They don't want in a year's time someone to say "Well here they are funding fresh water and here they are building a dam messing up this river". So, its rather ugly, but by taking money has actually raised the stakes on them. And that's the clever way of doing it. (Interviewee, WWF UK)

The above demonstrates how an organisational outcome benefits not only the organisation but also the environment through the change of organisational policies, hence demonstrates the *multidimensionality of change* and the *interfaces between organisational and social change*.

Indeed organisational genesis is a potential outcome of partnerships but the essential question to be asked *is whether companies (or NPOs) really want to change*. The *intention for change* appears to be important for change to be achieved not only as a process but also as an outcome. The type of change appears to depend on *the real will to change*:

> And that's, I mean that's essentially, the core of it is "do they really want to change?" Yeah, Nike could change a lot more than it has if it really wanted to. But I don't think there is a real will to change ... even companies with bad records can change. It just depends on how much they actually want to do it. (Interviewee, Greenpeace)

The above provides further evidence of the different types of change that can take place within organisations. It appears that what usually causes apprehension to society, and hence NPOs, *are the claims of organisational genesis* when indeed what takes place is only organisational kinesis of a sort. Hence the role of NPOs in partnerships is very important and they need to remain focused on their missions and remain clear about the aim of the relationship:

> They [NGOs] need to go in with their eyes wide-open. Naïve NGOs loose their objectivity. (Interviewee, Stop the Esso Campaign)

It seems that the radical NPOs continue to have doubts as to whether a collaborative approach is the best way of challenging businesses or achieving change in the behaviour of BUSs:

> I think it is really naïve to go into a partnership with business and don't think that they are not actually trying to influence what you are doing and put pressure on you, in terms of the results that come out. I think, you have to be unbelievably naïve to think that that doesn't happen. (Interviewee, FoE)

It appears that the experienced collaborative NPOs (e.g. WWF) only recently realised the risks in forging close relationships with BUSs and decided *to assume a more critical role even publicly similar to the radical NPOs*. Hence it appears that closer collaboration among the different NPOs, but also the exchange of information and facts, might increase openness and transparency in NPO-BUS partnerships, increase civil society's trust of organisations and institutions, and might also allow partnerships *to reclaim their societal role*.

5.7   Conclusion

The final empirical chapter discussed the partnership outcomes within both case studies under examination and the comparative interviews. The most valued organisational outcome for NPOs is the financial support they receive from BUS in the majority of cases, and for BUSs the HR opportunities they have access through the NPOs. The partnership process generates additional outcomes beyond the anticipated ones. The research indicated that a higher concentration of content organisational outcomes might be the result of more experienced organisations in forging relationships.

Based on the findings there is a clear difference between espousing a partnership approach in the relationship between NPOs and BUSs and claiming they have forged a social partnership. In the first case the relationship between the two organisations is characterised by higher levels of proximity over time, hence the process can generate more positive outcomes that encompass social outcomes due to the social role of the NPO. In a social partnership, however, the societal outcomes would appear as content outcomes as the intent for solving a social problem would entail a clear definition of the issue and a strategy from the outset.

Organisational genesis can be a borderline organisational/societal outcome of partnerships as changing the programmes or policies of either the profit organisation or NPO can accrue positive outcomes for society demonstrating the interface between organisational and social change. However, organisational kinesis (non-core changes) would not suffice as societal outcomes.

Collaborative NPOs need to assume a more critical role towards BUSs and increase their collaboration within their own sector in order to contribute and strengthen the societal dimension of partnerships.

Chapter 6
Reclaiming Responsibilities

6.1 Introduction

The final chapter discusses the main findings of this research within the literature in order to highlight the theoretical contributions of the study. Each empirical chapter answered the research questions within the three chronological stages of partnerships: formation, implementation and outcomes. Employing the chronological stages of partnerships and putting forward specific constructs for the study of each stage aimed to: (a) achieve a clear organisation of the theoretical themes and findings; and (b) allow for observations beyond any single stage. This chapter discusses the overarching themes of the book that extend beyond each stage of the partnership phenomenon.

6.2 A Holistic Framework for the Study of Social Partnerships

Researchers in CSSP have examined the different stages in partnerships by concentrating in one the three chronological stages: formation, implementation, outcomes or constructs within each stage. In this research data were collected for all three stages which allowed for a holistic examination of the partnership phenomenon with the same case study. The holistic framework that the study employed (see Fig. 6.1) allowed for observations on the level of the *constructs* (e.g. organisational characteristics, motives), *stages* (e.g. formation, implementation) but also developing *overarching themes that encompass all three stages* (e.g. change, conflict).

Examining all three stages in partnerships within the same case study provides a more in depth understanding of the phenomenon under study but also improves the comparisons across cases, contexts and types of partnerships. Furthermore it allows for the analysis to move to a deeper level in order to observe the overarching themes that are shared across the three stages either by their presence or absence. In the case of this research the first overarching theme was the absence of conflict which is discussed below.

M.M. Seitanidi, *The Politics of Partnerships: A Critical Examination of Nonprofit-Business Partnerships*, DOI 10.1007/978-90-481-8547-4_6,

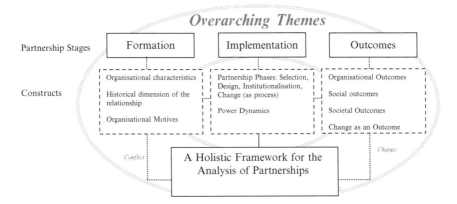

Fig. 6.1 A holistic framework for the analysis of partnerships

6.3 Overt Functional Conflict Deficit in NPO-BUS Partnerships

This section discusses the overarching theme of overt functional conflict (OFC) deficit by presenting the local conditions in NPO-BUS partnerships that are employed to organise the evidence.

The aim of this research was to study the phenomenon of NPO-BUS partnerships. Although the study examined collaborative NPOs in partnership with BUSs, it also included a number of critical NPOs such as Greenpeace, FoE, Partizans, Platform, Stop the Esso Campaign, International Textile Garment and Leather Workers' Federation, comprising significant exceptions. For methodological purposes they provided a counter-balance to the collaborative NPOs that participated in the study such as Earthwatch, The Prince's Trust, WWF, IPRA, A&B, WBCSD and CBI. However for the purposes of the research many more NPOs were researched although not formally interviewed. The reason that other NPOs were examined was to compare for example Rio Tinto's partnership with Earthwatch to the partnership the company has with Bird's Life, or to understand better the relationship between Deutsche Bank and the charities it supports. Hence the observations below, although they are primarily based on the interviews, they are grounded on a wider range of material consulted for this study.

NPO-BUS partnership is a widespread phenomenon in the UK which constitutes an important part of the broad view of politics (Scott and Marshall 2005; Calhoun 2002) as it legimitimises the joint decision-making in areas that were previously considered part of public policy (Waddock 1988). The new arrangements have been portrayed to represent the different perspectives of two economic sectors that previously embodied substantial differences including their ideas, values and objectives (Holzer 2001), goals (Kanter 1999) and characteristics (McFarlan 1999). Hence the joint decision-making appeared to signify the representation of different perspectives of individuals and organisations as they were allowed to express their opinions (Calhoun 2002) and participate in the governance of society within secluded spheres of expertise (environment, youth, and so forth).

As a result it would be expected to observe "conflict arising from the different values, issues or concerns of various groups" that in effect would constitute political behaviour within organisations (Hatch 1997:300). The evidence presented in this study suggests that when collaborative NPOs form partnerships with BUSs there is an overt functional conflict deficit. Overt functional conflict (OFC) in NPO-BUS partnerships can be defined as *the opportunities that appear during the interaction between the partner organisations for the expression of divergent opinions that would encourage the adaptation leading to change of perceptions, policies and actions. The OFC deficit refers to insufficient occurrences of such opportunities.*

Based on the model for inter-unit conflict of Hatch (1997:308–313) and the data the study examined, the book suggests a model to identify the possible sources of OFC deficit in NPO-BUS partnerships.

Figure 6.2 presents a model for identifying occurrences of OFC deficit under the chronological stages in NPO-BUS partnerships which consist of the local conditions of the partnership. Inverting the syllogism of Walton and Dutton (1969), the indicators represent what it is likely to be observed or experienced when there is an absence of conflict. Each one is discussed below within the research findings.

OVERT FUNCTIONAL CONFLCIT DEFICIT

Partnership Conditions ⟶ Observable Indicators of OFC Deficit

STAGE 1: Partnership formation
Organisational Characteristics   Similarities, commonalities, compatibilities
History of Relationship   Synagonistic relations, familiarity
Organisational Motives   Compatible motives

STAGE 2: Partnership Implementation
Task Interdependence   Non-reciprocal task performance related to the partnership
Status Incongruity   One way direction of authority
Jurisdictional Ambiguities   Clear delineation of responsibilities

STAGE 3: Partnership Outcomes
Outcome Imbalances   Prioritisation of organisational over societal outcomes

Fig. 6.2 Model for identifying sources of OFC deficit in NPO-BUS partnerships (Based on Hatch 1997)

6.3.1 Stage 1: Partnership Formation

The three constructs suggested for the study of partnerships under the stage of formation: organisational characteristics, the history of the relationship and partners' motives are discussed below. The structural characteristics suggested by Burger et al. (2004) are examined in order to assess which ones provide meaningful occurrences of similarity, compatibility and commonality between the partners, but also in order to present the organisational characteristics that this research contributes to the field. More specifically, one of the structural characteristics is the distinction between 'autonomy' and 'central control', referring to the management of the NPO, suggested by Burger et al. (2004), which does not provide an adequate difference between types of NPOs within this research. When a BUS formulates or implements a partnership it is usually executed through the headquarters of the NPO. Even in the case of Earthwatch, which comprises a network of national organisations, the autonomy and central control are both combined by the regional headquarters. What in fact appeared more important for the decision of BUSs were the *areas of operation of the NPO* which mirrored the areas of operation of the BUS partner, and represents a commonality and compatibility between the partners. Furthermore Berger et al. (2004) suggested the dichotomy between 'revenue generating' and 'non-revenue generating' NPOs. This characteristic was not examined by the research as it was not relevant. Within the case studies PT and Earthwatch each represent consecutively "big well established versus small, entrepreneurial" comprising another structural characteristic (Berger et al. 2004:79). They suggest that the 'brand equity' and 'name recognition' are attractive assets for BUSs usually associated with 'big and well established' NPOs, but also 'small entrepreneurial' NPOs are considered more 'flexible, eager and energetic'. Within the case studies examined these characteristics were present in both NPOs. What appeared to be important was the *industry sector the NPO was operating in* which comprised a common and central issue of interest for the partners. The size of the organisations was very different and when compared the BUS and NPO partners there was a clear asymmetry favouring the BUS. However both *status* and *size* are contributing factors to the *organisational confidence* of the NPO. There was a clear difference on the level of organisational confidence between the two partners which was the result of the 'funding dependency' that is frequent in NPOs (Leiter 2005:5) and consists of a type of coercive isomorphism (DiMaggio and Powell 1983). Hence in the case of WWF (a big well-established NPO) with equal organisational confidence as its partner, was able to return money to Lafarge in order to exercise pressure publicly. Similarly PT returned money to RBSG, assuming its independence when it disagreed with the bank, unlike Earthwatch which although it faced internal opposition for the formation of the partnership did not appear to be in a position to assume a confident stance and play a more critical role towards its partner. Consequently, the unequal organisational confidence produces through funding dependency coercive isomorphism which is a process that results in similarity (Leiter 2005; DiMaggio and Powell 1983). The similarity refers to the acceptance of convergent opinions with regard to the reasoning or justification of the partnership relationship under the resource constraints.

Another distinction referring to the mode of operation, which Berger et al. (2004:76–78) suggest, is: "programmatic versus grant-making" NPOs; in the first case a charity runs programmes for its constituents that are usually centred on focused aspects of the cause it serves (e.g. PT: social exclusion). The grant-making NPO raises money in order to support a number of causes or different programmes in a particular area of interest (e.g. Earthwatch: supports scientific research in conservation and biodiversity). Berger et al. (2004) posit that the association of a company with a grant-making NPO "can lessen the company's exposure to political issues within the cause community since funds are dispersed to a wider base". On the other hand, they suggest that there is a greater tendency for the BUS partner to micromanage the programmes of a programmatic rather than a grant-making NPO. There is evidence that some of the PT's employees felt that there were instances that the bank was perhaps 'micromanaging' aspects of the partnership by suggesting what the Trust should do. Earthwatch's self-portrait as 'a non-political, non-campaigning and non-confrontational organisation' represents a *non-threatening ideology* which is a very attractive non-structural characteristic for BUS. Although PT or other collaborative NPOs do not explicitly 'advertise' their political position, they share the same ideology which signifies the values and tactics they employ and which shape their organisational identity (Basler and Carmin 2002). Despite the different modes of operation (profit/nonprofit), the non-threatening ideology represents another point of compatibility between the collaborative NPOs and BUSs. Hence the ideology of collaborative NPOs appears to be compatible with the ideology of BUSs.

The final structural characteristic suggested by Berger et al. (2004) relates to the mission of the organisations regarding the approach towards BUSs ranging from 'inherent cross-sector collaboration' versus 'traditional NPOs'. The first type refers to the trend of nonprofits that "are founded on the premise of fostering corporate partnerships" (Berger et al. 2004:80). The traditional NPOs, on the other hand, according to Berger et al., employ the relation with BUSs as one of their strategies for income generation. The above is an important observation and this research confirms a similar trend in the UK[1] that confirms the increase of collaborative relations but also the inclusion of *collaborative intent in the missions of NPOs* in fostering partnerships with BUSs.

Based on this research another important characteristic is the *domination of network culture* that induces similarities between the partners and it is produced by the increased interactions within clusters of organisational networks that share similar values and perceptions about the role of partnerships with BUSs. The case of Earthwatch provided evidence of how the organisation consulted only with like-minded organisations, which was further verified by the more recent attempts of organisations (e.g. Oxfam and Greenpeace) to broaden their collaborations beyond

---

[1]The research conducted by Berger et al. (2004) took place in the context of the US involving 26 organisations.

their area of expertise. DiMaggio and Powell (1983) suggest that the interaction within a collection of organisations with substantial relations can be "between equals or unequals; it may be competitive or collaborative; it may or may not involve the exchange of resources" (Leiter 2005:3) and it produces isomorphism. The result is the imitation of the 'successful' model of leading organisations in developing relationships with 'successful' BUSs which is regarded as *mimetic isomorphism* (DiMaggio and Powell 1983).

Figure 6.3 presents a picture of NPO-BUS partnerships between collaborative NPOs and BUSs. The red circles represent BUSs and the green NPOs. Each line represents a partnership relationship. This is a partial picture of a broad social network across the two sectors highlighting the close link between BUS and NPOs that induces mimetic isomorphism within partnerships as the relationship patterns are carried through from one relationship to the next. It also portrays how closed the network structure is, suggesting the network pressures that are exercised to all its constituent parts in ascribing to the predominant institutional culture and tactics of operation within the network. Another important observation is that mimetic isomorphism is spread across different industries (mining, oil and gas, tobacco, insurance and banking-financial) through environmental partnerships with collaborative NPOs, as observed in the figure. The collaborative environmental NPOs

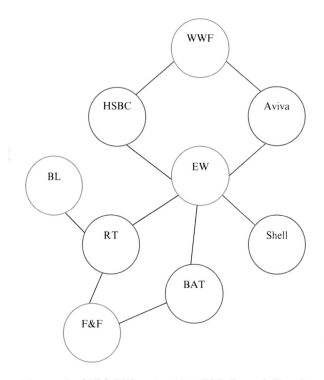

Fig. 6.3 A social network of NPO-BUS partnerships (F&F: Fauna & Flora International; BL: Birdlife; BAT: British American Tobacco; HSBC Bank)

provide the *legitimacy connectors* across the different industries. However, what appears to be the case, based on this research, is that the *NPOs do not exercise their network power in order to strategically and collectively pursue their missions and attempt to push BUSs to change* (with the exception of the WWF-Lafarge case WWF); instead the financial dependency seems to capture their ability to express difference and criticism which in effect increases the overt functional conflict deficit.

Despite the differences of NPOs in a number of characteristics such as sector, size, status, revenues, mode and scope of operations, there are overarching characteristics that contribute to the lack of difference and disagreement in collaborative NPOs. The structural characteristics that the research contributes suggest they determine an important role in the partnership including: *the areas of operation of NPOs, the industry sector of NPOs, the collaborative intent in the missions of NPOs* and *the domination of network culture.* The non-structural characteristics include the *organisational confidence,* and *the non-threatening ideology of NPOs.* Increasingly collaborative NPOs foster an inherent political position that accepts the prominent role of BUSs in the global debates, hence in order to broaden their agendas and avoid issue isolation or safeguard much-needed financial resources they engage in developing mutually beneficial partnerships. This appears to result in a *gradual convergence between BUSs and collaborative NPOs.* The collaborative mode of approach towards BUSs assumes a 'sensemaking lens' (Carmin and Basler 2002) that shapes the decision-making process, and as a result, shapes actions. Hence the similarities, commonalities and compatibilities across the partners do not allow for difference or disagreement and in effect they contribute to the overt functional conflict deficit.

Similarly the *historical dimension* examined across the partner organisations revealed *synagonistic relationships* over a number of years that produced gradual *familiarity,* which increased further due the partnership relationship. Indeed as suggested by Bennett and Savani (2004:192), activities such as team-working "create common perspectives on problems and how they might be resolved" hence the enhanced familiarity decreases the potential for functional conflict. Furthermore the absence of conflict "can have negative consequences such as group think, poor decision making, apathy and stagnation" (Hatch 1997:304). Indeed *complacency* was an occurrence within the Earthwatch-Rio Tinto partnership observed by the actors that constituted a problem within the partnership relationship.

Furthermore the comparison of the partners' motives revealed a high degree of *compatibility* with a *prioritisation of the organisational needs for both organisations.* The *homogenous relationship* that is formed between collaborative NPOs and BUSs gradually removes threats and challenges from within the relationship. The combination of reciprocity as a motive (Oliver 1990) emphasises *the collaboration and convergence of interests* due to the pursuit of mutually beneficial outcomes along with the inexperience to balance the asymmetrical power distribution, that originally functions as a motive, results in decreasing the potential for overt functional conflict. The research suggested the need for external support in order to balance the power asymmetry through the involvement of other NPOs due to their expertise.

6.3.2   Stage 2: Partnership Implementation

Within the second stage of partnership implementation the task interdependence (Hatch 1997) refers to the extent the partners are working together in order to implement a task. In social partnerships, partners work towards a common goal (e.g. developing a new product, programme or policy) that requires continuous interaction among the partners. Hence in this case a number of tasks would be performed involving pooled, sequential or reciprocal task interdependence (Hatch 1997). However, in the partnerships that were examined this did not appear to be the case. Although collaboration was increased among the partners, however, the interaction did not appear to allow for opportunities of interdependence. Although there was a virtual team in the case of the PT-RBSG, their work did not appear to be dependent on each other as the programmes that were developed were delivered only by PT. The same applies to the Earthwatch-Rio Tinto partnership. Hence the lack of task interdependence did not allow for opportunities of functional conflict to arise during the implementation stage, in effect contributing to the OFC deficit.

Status incongruity refers to the difference of statuses across the partner organisations. Hatch suggests (1997:312) that the "imbalance of status is not problematic as long as higher status groups influence lower status groups, however if lower status groups must initiate activities or otherwise directly influence higher status groups, then conflict is more likely to occur". The BUS organisations in both partnerships clearly represented the case of higher status groups in the perceptions of the organisational actors. This is evident through the expression of opinions about each other and the level of perceived financial dependency. However, in the case of the WWF-Aviva partnership, where there was no financial dependency, the status imbalance appeared to be decreased. Similarly, in the WWF-Lafarge partnership due to the initiation of a challenging activity by the NPO, OFC was publicly expressed balancing the power dynamics across the partners.

When the responsibilities are unclear between the organisations conflict can be produced due to the inability to assign blame or credit to either organisation, which refers to the jurisdictional ambiguities (Hatch 1997). Collaborative NPOs appear to either refuse (e.g. PT) or are unable (e.g. Earthwatch) to form any evaluative statement about their partners with regard to their areas of expertise: *Earthwatch was not able to accept the responsibility of playing a role of assessing the biodiversity policy of Rio Tinto but as a result implicitly legitimised the company's activities without any in-depth assessment.* Similarly *PT implicitly refused the responsibility to challenge RBSG with regard to its policies towards the young people the Trust is assisting.* Hence the possibility of any form of OFC is captured in the politics of the partnership deriving from their need to meet their organisational needs such as funding, reputation, image within certain groups and networks.

6.3.3   Stage 3: Partnership Outcomes

Finally, the prioritisation of the organisational over societal outcomes by both BUS and NPO partners, discussed extensively in the previous chapter, does not allow for outcome imbalances to occur, and further, to be negotiated. If indeed one of the organisations chose to prioritise the societal outcomes through the relationship, which could generate a certain amount of conflict- not necessarily dysfunctional- then instances of OFC might occur. Prioritising the *societal outcomes* would allow both organisations to reclaim their responsibilities as socially embedded organisations. Although the government can be blamed for the decrease of available resources to the nonprofit sector which resulted in the prioritisation of their financial needs, this challenge can be dealt with in other ways, such as mergers between NPOs in areas such as conservation, social exclusiont and so forth. This would allow NPOs to *reclaim their responsibilities* not only across sectors but also within their own sector and areas of expertise.

Summarising the above, the local conditions of partnerships between collaborative NPOs and BUSs throughout the stages of the relationship present an OFC deficit. However, manifestations of covert conflict through examples such as the 'silent' dislike of one-way reporting, the existence of a sub-culture with divergent opinions of the predominant status quo within NPOs (e.g. the initial resistance of Earthwatch's stake-holders to form a partnership with Rio Tinto), offer evidence of the importance of the *informal side of the organisational interactions highlighting the existence of a covert side of conflict.* The covert side of conflict presents an opportunity for NPOs to embrace the diversity of opinions by institutionalising the expression of difference which in effect would introduce formalisation processes aiming to change the covert conflict into overt. Furthermore, it might assist in releasing the pressure that is exercised by the informal network culture and forces organisations to mimetic practices. The formalisation of processes can assist collaborative NPOs in reclaiming their societal responsibilities within the partnership relationship.

The next section discusses change and suggests a link between the OFC deficit and organisational genesis, a type of organisational change.

6.4   Change as Process and Outcome in NPO-BUS Partnerships

A second overarching theme within the NPO-BUS partnerships is change which appeared throughout the partnership stages. The study aimed to address the previous criticism of Pettigrew (2000:243) that research on organisational change is "largely acontextual, ahistorical and aprocessual". Firstly it studied change within NPO-BUS partnerships which provided the context for studying change as a relationship outcome on the personal, organisational and societal levels. It concluded that there is a profound difference between organisational genesis and kinesis in all different levels of change. Also it introduced the historical dimension to the phenomenon under study by adapting the Barzelay's (2001) model in order to account for the relationship evolution and conclude on the historically synagonistic relationship that exists between the collaborative NPOs and BUSs as a contributing factor for the OFC deficit. Finally, the research incorporated change as part of the partnership process, as observed within the case studies, suggesting that intentional change can be the final stage of the partnership implementation, if a strategy is in place and can result in organisational genesis or kinesis and in effect content or processual outcomes. The above confirms the conceptualisation of change as an interaction field (Lovelace et al. 2001) where "content and action are inseparable" (Pettigrew et al. 2001:6997). NPO-BUS partnerships provide a multi-sector and multi-dimensional context within which the *intention* of change (intentional/unintentional), the *level* (organisational, personal, and societal), the *type* (content/processual) and the *effect* (genesis/kinesis) can be observed.

The findings suggest that in the case of partnerships between collaborative NPOs and BUSs there is a link between overt functional conflict deficit and change. When the intention for change is evident in both partners (in any combination: e.g. the BUS or the NPO facilitating the change process within its partner) then organisational genesis can be achieved. However, when the intention for change is only evident in one of the partners then the overt functional conflict deficit decreases the potential for organisational genesis. The OFC deficit is linked with the culture of collaborative NPOs. Schwartz and Davis (1981:35) suggest that: "culture is capable of blunting or significantly altering the intended impact of even well-thought-out changes in an organisation". As pointed out by Senior and Fleming (2006:173): "organisational culture comes in many forms, and therefore, can be more or less supportive of change".

Based on the data of this research, in the case of collaborative NPOs partnering with BUSs the intent and process of facilitating change within their BUS partner is not yet well-articulated. The reasons for this follow from: (a) the organisational characteristics of the collaborative NPOs; and (b) a number of issues within the relationship management process that consists of: (1) the lack of experience in managing conflict internally and externally; (2) the conflict that remains captured and hence it is demonstrated only as covert within the relationship; (3) the lack of institutionalised modes for the expression of plurality of opinion; (4) the financial dependence of the NPOs from their BUS partners.

On the other hand, it appears that there is a potential of externalising the conflict in NPOs if there is no financial dependency or if it is removed (e.g. WWF-Lafarge partnership), or if a shift in the mission of the organisation takes place with regard to the role and responsibilities in forming close relations with BUSs and if the societal outcomes are prioritised over organisational outcomes.

This section suggested that change can be both a process and outcome within the context of NPO-BUS partnerships and there is a relationship between organisational genesis, a type of change and OFC deficit: organisational genesis is possible only when the intent for change is evident in both partners.

The next section will discuss the difference between the partnership form and partnership approach.

6.5 Partnership Form Versus Partnership Approach

Austin (2000), and earlier Waddock (1988), suggested that the added value of the partnership should not only be delivered to either or both partners but also to society. The partners should answer the question of "how society is better off because of their joining resources and efforts" (Austin 2000:88). The previous chapter on partnership outcomes suggested a triple distinction between organisational, social and societal outcomes. Positive social outcomes in NPO-BUS partnerships refer to the benefits that accrue as a result of the work of the NPO through the support of BUS; *Societal outcomes* refer to the benefits that accrue as a result of the *combined efforts* of the two partners and which are *delivered externally to society* and not to the partner organisations. In effect societal outcomes can deliver the '1 + 1 = 3' effect which refers to the 'over and above the parts' outcome. In other words the combination of the capabilities of the partners to a new combined capability over and above each compartmentalised reality that has the potential to externalise the benefits to society, produce change and offer solutions to social problems. As such partners can achieve unity through diversity and progress to a unity of goals when aiming to reach the social sphere. The difference between social and societal outcomes distinguishes the *social partnerships form* from the *partnership approach*. Social partnerships, as defined by Waddock (1988), are a different genre of interaction between the sectors and point towards an alternative form of governance (Moon 2004; Rowe and Devanney 2003; Miller and Ahmad 2000), a non-regulated one where all sectors of society are trying to address together a multitude of pressing social problems. According to Nelson and Zadek (2000:55) "governance is today increasingly about the roles, responsibilities, accountabilities and capabilities of different levels of government (local, national, regional and global) and actors or sectors in society (public, business and civil society organisations)". In the case of the *partnership approach* organisations within the same economic sector or across different economic sectors join forces in order to develop a competitive advantage, exchange resources or in general support the work of each other. The partnership approach can deliver more intangible and processual benefits due to the proximity and familiarity between the partners, unlike other approaches such as transactional, or philanthropic. Positive societal outcomes can accrue from organisations who work in partnership but they predominantly will not: (a) be intentional; (b) mutually address a social problem, hence they will not be regarded as an alternative form of *societal governance*. In order for a relationship to be characterised as *a social partnership* societal outcomes need to be prioritised clearly from the outset. In the case of a partnership approach organisational rather than societal needs are prioritised within a partnership relationship. Hence increasingly the previously termed 'social partnerships' are referred as 'strategic partnerships', as the focus has shifted from serving society to satisfying organisational needs. It appears that in most cases *the politics of the NPO-BUS partnerships fail the societal dimension of partnerships by capturing the potential of organisational genesis (or core changes) within the convergence of need rather than divergence of missions.*

Another aspect of difference between a social partnership and 'working in partnership' arises from the expectations that a social partnership raises. For example, when an environmental NPO suggests that it works in partnership with a BUS and that it aims to 'change' its policies or practices without in fact being in a position to deliver such a change, this creates a *perception gap* within the organisation about its role, across different organisations and beyond (i.e. within society) about what a NPO can deliver and what it claims that it is delivering. Perception gaps contribute to the schism between rhetoric and reality with regard to the roles and responsibilities of each sector and its abilities. One of the reasons behind the difference between the rhetoric and reality of partnerships is the level of social institutionalisation of the partnership and its aims. As the findings of the research suggest the *partnership institutionalisation* is possible and it was suggested that the extent of institutionalisation can be observed through the attainment of processual outcomes. Moreover, if the effect of the outcomes is to extend beyond the organisational level then perhaps the partnership needs to be institutionalised within society. In other words, the partnership relationship needs to be embedded not only within a single nonprofit organisation but rather it must involve a wide range of diverse NPOs that would be able to offer advice and contribute expertise and diversity of opinion in the decision-making process of the focal NPO.

The previous sections discussed the overarching findings of the book, which are summarised below.

6.6   Conclusion

The chapter discussed the overarching partnership outcomes within both case studies by putting forward a holistic framework for the analysis of social partnerships which allowed for observations beyond any single stage of partnerships. The first overarching theme that was shared across all three stages and across both case studies was the OFC deficit referring to the insufficient opportunities for the expression of divergent opinions that would lead to fundamental changes. As such, change was the second overarching theme that was observed as a process and outcome. The connection between the OFC deficit and change lies in the presence of intention for change. If the intention for change is evident in both partners then change can take place if accompanied by a strategy. However, if change is evident in one of the partners then the OFC can increase the possibility of organisational genesis; equally the OFC deficit acts to decrease the potential of fundamental type of change. Finally, a clear distinction was put forward between the partnership as a form and as an approach. In the first case a social partnership needs to prioritise from the outset societal outcomes that will derive from the relationship. In the case of a partnership approach organisational rather than societal needs are prioritised and the partnership primarily indicates the proximity of collaboration between two organisations.

Epilogue: Beyond Boundaries

Organisations are embedded within institutional environments bounded by divides such as confrontational and collaborative NPOs with opposing perceptions of how change in BUS can be achieved. Containing dilemmas within networks of organisations that share the same understandings and values captures the conflict that previously existed in the social domain among NPOs and BUSs. The institutional logic that exists within networks of organisations nurtures the *typical voices* which empower the existing conceptions with regard to which actions are acceptable, legitimate and do not pose threats to the status quo. This research suggests that when collaborative NPOs are partnering with BUSs there is an overt functional conflict deficit observed in the partnership stages of: formation, implementation and outcomes, based on: (a) the organisational characteristics of the collaborative NPOs; (b) the lack of experience in managing conflict internally and externally; (c) the capture of conflict that remains covert within the relationship; (d) the lack of institutionalised modes for the expression of plurality of opinion; (e) the perceived financial dependence of NPOs from their BUS partners.

The research suggests that there is the potential for externalising the conflict if NPOs do not perceive that they are financially dependent on their BUS partner, there is a shift in the mission of the organisation with regard to the role and responsibilities in forming close relations with BUSs, and/or if societal outcomes are prioritised within the relationship. The above can be supported by involving diverse NPOs within the relationship in order to increase the external legitimacy of the relationship and to offer diverse type of expertise beyond the single NPO and the network of NPOs in which it operates (Seitanidi, and Lindgreen, 2009).

As noted by Hatch and Schultz (1997:356), "one of the primary challenges faced by contemporary organisations stems from the breakdown of the boundary between their internal and external aspects. Previously, organisations could disconnect their internal functioning from their external relations in the environment because there were few contacts between insiders and outsiders". The breakdown of the boundaries is a global phenomenon associated with globalisation and the communications revolution which allows for speedy and low-cost communication across and between actors, organisations, states, and ideas either in a formal and structured manner, or in an informal manner. According to Hatch and Schultz (1997:356) there are " increasing levels of interaction between organisational members and suppliers, customers, regulators and other environmental actors, and the multiple roles of organisational members who often act both as 'insiders' (i.e. as employees) and as 'outsiders' (e.g. consumers, community members and/or members of special interest groups". These changes imply there is a gradual collapse of boundaries between different levels of reality, such as internal/external, formal/informal, local/global, profit/nonprofit/government suggesting a need for an increased ability for individuals and organisations to manage simultaneous multiple roles across different levels. Particularly in the case of collaborative NPOs their characteristics further contribute to the proximity of perceptions with BUSs due to the increased historical interactions across the sectors. The above also contribute to convergence of perceptions across networks of organisations from different economic sectors. The increased interactions push the previously existing boundaries and demand a subtler 'sensitivity' and awareness towards the possible outcomes of actions. The multiple roles and relaxation of boundaries contribute to a higher level of isomorphism which creates pressures to organisations that need to abide on the one hand by the pre-existing institutionalised norms and on the other hand must compete for survival within the changing rules. In the case of partnerships, the pressures that NPOs experience in their efforts to resolve dilemmas and paradoxes remain contained and hence unresolved within the networks of like-minded organisations. Complex response processes in organisations (Stacey, 2001) prioritise processes and interaction in localities in which boundaries become meaningless because "process is essentially about movement, which is both spatial and temporal at the same time but not boundaried" (ibid: 168). Hence, in partnerships with collaborative NPOs and BUSs there is an increased need for reclaiming the difference that stems from the distinctive responsibilities the profit and nonprofit sectors represent in order to overcome the sectoral ambiguity (Lewis, 1998) that appears to emerge. Another aspect that is important in the emergence of responsibilities is time. As issues surface responses do not emerge simultaneously, due to the perceived divide of responsibilities of the sectors. Moving away from uni-dimensional responsibilities that are either re-active or pro-active to multi-dimensional responsibilities can demonstrate the ability of a social system to adapt at the same time an issue emerges (Seitanidi, 2008). "Adaptive responsibilities demonstrate an adaptation to an emergent issue that extends beyond a single dimension of responsibility (tri-mensional reality) in order to offer solution to problems that require fundamental change. The responsibilities become adaptive when the entity is able to transcend beyond its self-centred reality" (Seitanidi, 2008: 61).

Following the above descriptions of today's organisational reality it seems rather difficult to hold the assumption that organisations are distinctive based on the economic sector they represent and maintain the original characteristics that each one's sector was predominantly ascribed with. Co-operation appears to be a default property of the endemic boundary-less institutional environment where understandings are shared for the purpose of organisational survival. The failure of the form of partnership to focus 'openly and clearly' (Caplan, 2003:34) on the societal outcomes can contribute not only to the disappointment of the partners (ibid) but more importantly to decreasing public trust in the ability of the partnership institution to deliver the promised 'social goods'. The new institutional terrain may require a better appreciation of the differences and similarities across organisations and hence a redefinition of the aims and responsibilities of the economic sectors and subsectors in order to accommodate the new era of increased interaction and its effects. A possible future scenario is that the distinction of profit and nonprofit organisational form will become obsolete as the signification will not denote enough difference in the consequences. Hence, the increasing appropriation of organisational form will require stronger mission statements from organisations that will clearly prioritise sustainable societal over financial outcomes, being prepared to sacrifice profit as a non-important outcome for the purpose of serving a fundamental social good. A radical shift in the prioritisation and desirability of outcomes by citizens is inevitable and will safeguard sustainable solutions to social problems while assisting in the realignment of the political, social and economic dimensions of social life. As the state inevitably has lost its ethical monopoly to provide society with 'social goods' the participation of citizens in political decisions is taking place through their role as active citizens rather than as detached voters by exercising direct or indirect governance. The micro-associational domain of partnerships in which the roles and responsibilities are negotiated is a terrain that requires new skills by individuals. In this terrain intelligent decision making will take place developing a new appreciation of what constitutes legitimate and accountable social life prioritising the notion of 'we' before the demands of 'I' in order to safeguard that there will be a world where the individual would wish to live in tomorrow.

Appendices

Appendix 1

Material consulted for the in-depth case studies: (A) Earthwatch-Rio Tinto Partnerships and (B) The Prince's Trust-Royal Bank of Scotland Partnership

A. Case Study Earthwatch-Rio Tinto Partnership

Earthwatch Organisational Literature

Case Studies in Business & Biodiversity. Published by Earthwatch on behalf of DETR. March 2000

Business & Biodiversity. A guide for UK based companies operating internationally

Earthwatch. Business & Biodiversity. Sire Biodiversity Action Plans. A guide to managing biodiversity on your site

Earthwatch and Sustainable Development

Earthwatch Annual Report 2001

Earthwatch Annual Report 2001/02

Earthwatch Annual Report 2003

Earthwatch. Implementing Corporate Environmental Responsibility 2002–2003

Earthwatch. The news magazine for fellows

Earthwatch. Everything you need to know about joining an Earthwatch team. Booking conditions

Earthwatch. Life-changing opportunities for you and the environment. A selection for our research projects for volunteers 2003

Earthwatch. British American Tobacco Employee Fellowships 2003

Earthwatch Institute. 2003–2004 Research and Exploration

Earthwatch Institute. Business & Biodiversity. The Handbook for Corporate Action

Internal Documents: Reports, Papers, Business Plan

Report to Rio Tinto on the Partnership between Rio Tinto and Earthwatch Institute (Europe) between January 1999 to March 2001. Earthwatch, April 2001

Earthwatch Business Plan 2002–2003. Earthwatch, October 2002

Report to Rio Tinto on the 2002 programme of activities. Earthwatch, September 2003

Report to Rio Tinto on 2002 Programme of Activities. Earthwatch February 2003

The Globalisation of Corporate-NGO Partnerships. Case Study

Report to Rio Tinto on the 2002 Programme of activities

Earthwatch paper on: Can looking at risks and benefits help define partnership within the Rio Tinto context? May 2003

Earthwatch Institute (Europe) Programme Plan: January 2002–December 2004

Earthwatch. Project Better World. Earthwatch Fellowship Programme 2003

Rio Tinto Organisational Literature

Rio Tinto & Earthwatch Institute. Building Capacity for scientific research and biodiversity conservation. October 2001

Rio Tinto. Review. September 2002

Rio Tinto. Review. December 2002

Rio Tinto. Review. March 2003

Rio Tinto. Review. June 2003

Rio Tinto & Earthwatch Institute 2003. Global Employee Fellowship Programme

Rio Tinto. Education with Communities. Rio Tinto's support of educational initiatives in Australia and New Zealand

Rio Tinto Business with Communities Programme. Corporate Citizenship in Australia

Rio Tinto. Business with Communities Programme. Partnering

Rio Tinto. Topic Paper. Grouping for net strength. November 2000.

Rio Tinto. The way we work. Our statement of business practise

Rio Tinto. The way we work. Our statement of business practise. April 2003

Rio Tinto. Sustainable Development. Rio Tinto's contribution on the Rio decade

Rio Tinto. Data Book 2001

Rio Tinto. Project Platypus. Using Science to Plan

Rio Tinto-BirdLife International Programme

Rio Tinto. Annual Review 2001

Rio Tinto. Annual Report and Financial Statements

Rio Tinto. Annual Report 2002

Rio Tinto Annual Report 2003

Rio Tinto-Deakin University. Proceedings of the Second National Conference on Corporate Citizenship. November 16–17, 2000. Strategic Corporate Citizenship

Professor David Birch. Doing Business in New Ways. The Theory and Practise of Strategic Corporate Citizenship with specific reference to Rio Tinto's community Partnerships. Corporate Citizenship Unit, Deakin University

Professor David Birch. Rio Tinto's Business with Communities Programme: An Enabling Environment for embedding Corporate social responsibility visions into core business

Corporate Citizenship Unit, Deakin University

Exchange of Emails

Public Presentation of Community Relations Manager

Rio Tinto Video Tapes (4)

Rio Tinto's website 2002, 2003 and 2004

Rio Tinto 2001. Social and environmental highlights

Rio Tinto 2002. Social and environmental review highlights

Rio Tinto. A guide to our corporate identity

Rio Tinto. Minerals and metals for the world

Rio Tinto. Global Business, local neighbour. Community Relations 2001

Rio Tinto and the Australian Science Olympiads. A Rio Tinto Business with Communities programme

Rio Tinto. Foundation for a Sustainable Minerals Industry

Rio Tinto and the Corporate Citizenship Research Unity, Deakin University

Pacific Coal. Koala Venture. A Partnership between the University of Queensland and Blair Athol Coal

Rio Tinto and the WWF Frogs Programme. Partnering

B. Case Study The Prince's Trust-Royal Bank of Scotland Partnership

The Prince's Trust Organisational Literature

The Prince's Trust Annual Report for the Year Ended 31st March 2004. Incorporated by Royal Character

The Prince' Trust. Our Brand

The Prince's Trust. Breaking Barriers? Reaching the hardest to reach. August 2003

The Prince's Trust. 25 years. 2001

The Prince's Trust. Good ideas wanted. Start-up in business with The Prince's Trust

The Prince's Trust. Get out of your box. Team-up with The Prince's Trust

The Prince's Trust. Helps get young people's lives working. Invest in futures.

The Prince's Trust. Together we can help more young people. Look beyond the label

The Prince's Trust website 2002, 2003, 2004 and 2005

The Prince's Trust Annual Review 2001/02. Consequences

The Prince's Trust. The Business (magazine). Winter 2003

Internal Documents: Reports, Papers, Business Plan

The Prince's Trust and the Royal Bank of Scotland Group Business Awards for 2002/3. Internal documents related to the awards' process of selection

Royal Bank of Scotland Group/The Prince's Trust Account Plan, 2002–2003

A number of press cuttings with regard to the PT-RBSG partnership from local and national newspapers and magazines

Communications Key Documents. November 2003

The Prince's Trust and the Royal Bank of Scotland Group. Route 14–25. Interim Report. April 20003

The Prince's Trust and the Royal Bank of Scotland Group. Route 14–25. Review Document 2002/2003

The Prince's Trust. New Projects Bulletin

Senior Project Manager's Speech at the Institute of Fundraising, July 2003

Royal Bank of Scotland Organisational Literature

Royal Bank of Scotland Group, 2003. Annual Reports and Accounts. Make it Happen

Royal Bank of Scotland Group, 2002. Annual Reports and Accounts. Make it Happen

Royal Bank of Scotland Group, 2001. Annual Reports and Accounts. Make it Happen

Royal Bank of Scotland Group, 2000. Delivering on our promises

Royal Bank of Scotland Group, 2001. Community and Environment Report. Make it happen

RBSG. Your Magazine. September 2000

RBSG. Your Magazine. July 2002

RBSG. Your Magazine. September and March 2003

Royal Bank of Scotland's website 2002, 2003, 2004 and 2005

The Royal Bank of Scotland Group. The RBSG in the community

Internal Documents: Reports, Papers, Business Plan

Royal Bank of Scotland's intranet

Develop with the The Prince's Trust. Guidance, Referral and Nomination Process. April 2002: Version 3; September 2003: Version 4

Appendix 2

The International Classification of Nonprofit Organisations (ICNPO) provided by Salamon and Anheier (1997:70–74).

Group 1: Culture and the Arts

1 100. Culture and Arts

- Media and communications
- Visual arts, architecture, ceramic arts
- Performing arts
- Historical, literacy and humanistic societies
- Museums
- Zoos and aquariums
- Multipurpose culture and arts organisations
- Support and service organisations, auxiliaries, councils, standard setting and governance organisations
- Culture and arts organisations not elsewhere classified

1 200. Recreation

- Sports clubs
- Recreation/pleasure or social clubs
- Multipurpose recreational organisations
- Support and service organisations, auxiliaries, councils, standard setting and governance organisations
- Recreational organisations not elsewhere classified

1 300 Service Clubs

- Service clubs
- Multipurpose service clubs

- Support and service organisations, auxiliaries, councils, standard setting and governance organisations
- Service clubs not elsewhere classified

Group 2: Education and Research

2 100 Primary and Secondary Education

- Elementary, primary and secondary education

2 200 Higher Education

- Higher education (university level)

2 300 Other Education

- Vocational/technical schools
- Adult/continuing education
- Multipurpose educational organisations
- Support and service
- Organisations, auxiliaries, councils, standard setting and governance organisations
- Education organisations not elsewhere classified

2 400 Research

- Medical research
- Science and technology
- Social sciences, policy studies
- Multiple research organisations
- Support and service
- Organisations, auxiliaries, councils, standard setting and governance organisations
- Research organisations not elsewhere classified

Group 3: Health

3 100 Hospitals and Rehabilitation

- Hospitals
- Rehabilitation hospitals

3 200 Nursing Homes

- Nursing homes

3 300 Mental Health and Crisis Intervention

- Psychiatric hospitals
- Mental health treatment
- Crisis intervention
- Multipurpose mental health organisations
- Organisations, auxiliaries, councils, standard setting and governance organisations
- Mental health organisations not elsewhere classified

3 400 Other Health Services

- Public health and wellness education
- Health treatment, primarily outpatient
- Rehabilitative medical services
- Emergency medical services
- Multipurpose health service organisations
- Organisations, auxiliaries, councils, standard setting and governance organisations
- Health service organisations not elsewhere classified

Group 4: Social Services

4 100 Social Services

- Child welfare, child services, day care
- Youth services and youth welfare
- Family services
- Services for the handicapped
- Services for the elderly
- Self-help and other personal social services
- Multi organisations, auxiliaries, councils, standard setting and governance organisations purpose social service organisations
- Social service organisations not elsewhere classified

4 200 Emergency and Relief

- Disaster/emergency prevention, relief and control
- Temporary shelters
- Refugee assistance
- Multipurpose emergency and refugee assistance

4 300 Income Support and Maintenance

- Income support and maintenance
- Material assistance
- Multiple income support and maintenance organisations
- Organisations, auxiliaries, councils, standard setting and governance organisations
- Income support and maintenance organisations not elsewhere classified

Group 5: Environment

5 100 Environment

- Pollution abatement and control
- Natural resources conservation and protection
- Environmental beautification and open spaces
- Multipurpose environmental organisations
- Organisations, auxiliaries, councils, standard setting and governance organisations
- Environmental organisations not elsewhere classified

BM1.9.2   5 200 Animals

- Animal protection and welfare
- Wildlife preservation and protection
- Veterinary services
- Multipurpose animal services organisations
- Organisations, auxiliaries, councils, standard setting and governance organisations
- Animal-related organisations not elsewhere classified

Group 6: Development and Housing

6 100 Economic, Social and Community Development

- Community and neighbourhood organisations
- Economic development
- Social development
- Multipurpose economic, social and community development organisations
- Organisations, auxiliaries, councils, standard setting and governance organisations
- Economic, social and community development organisations not elsewhere classified

6 200 Housing

- Housing associations
- Housing assistance
- Multipurpose housing organisations
- Organisations, auxiliaries, councils, standard setting and governance organisations
- Housing organisations not elsewhere classified

6 300 Employment and Training

- Job training programmes
- Vocational counselling and guidance
- Vocational rehabilitation and sheltered workshops
- Multipurpose employment and training organisations
- Organisations, auxiliaries, councils, standard setting and governance organisations
- Employment and training organisations not elsewhere classified

Group 7: Law, Advocacy and Politics

7 100 Civic and Advocacy Organisations

- Civic associations
- Advocacy organisations
- Civil rights associations
- Ethnic associations

- Multipurpose civil and advocacy organisations
- Organisations, auxiliaries, councils, standard setting and governance organisations
- Civic and advocacy organisations not elsewhere classified

7 200 Law and Legal Services

- Legal services
- Crime prevention and public safety
- Rehabilitation of offenders
- Victim support
- Consumer protection associations
- Multipurpose law and legal service organisations
- Organisations, auxiliaries, councils, standard setting and governance organisations
- Law and legal organisations not elsewhere classified

7 300 Political Organisations

- Political parties
- Political action committees
- Multipurpose political organisations
- Organisations, auxiliaries, councils, standard setting and governance organisations
- Political organisations not elsewhere classified

Group 8: Philanthropic Intermediaries and Voluntarism Promotion

8 100 Philanthropic Intermediaries

- Grant-making foundations
- Voluntarism promotion and support
- Fundraising intermediaries
- Multipurpose philanthropic intermediaries and voluntarism organisations
- Organisations, auxiliaries, councils, standard setting and governance organisations
- Philanthropic intermediary organisations not elsewhere classified

Group 9: International Activities

9 100 International Activities

- Exchange/friendship/cultural programmes
- Development assistance associations
- International disaster and relief organisations
- International human rights and peace organisations
- Organisations, auxiliaries, councils, standard setting and governance organisations
- International organisations not elsewhere classified

Group 10: Religion

10 100 Religious Congregations and Associations

- Protestant churches
- Catholic churches
- Jewish synagogues
- Hindu temples
- Shinto shrines
- Muslim mosques
- Multipurpose religious organisations
- Associations of congregations
- Organisations, auxiliaries, councils, standard setting and governance organisations
- Religious organisations not elsewhere classified

Group 11: Business and Professional Associations, Unions

11 100 Business and Professional Associations, Unions

- Business associations
- Professional associations
- Labour unions
- Multipurpose business, professional associations and unions
- Organisations, auxiliaries, councils, standard setting and governance organisations
- Business, professional associations and unions not elsewhere classified

Group 12: (Not Elsewhere Classified)

12 100 NEC

Appendix 3

Earthwatch-Rio Tinto Partnership Content

The partnership between Earthwatch Institute Europe[1] and Rio Tinto was the first global partnership for both organisations. The partnership has several components:

1. Rio Tinto is a member of Earthwatch's prestigious Corporate Environmental Responsibility Group (CERG), which functions as an informal environmental consultancy for the members providing them with feedback and consultation on their policy and biodiversity programmes. It also functions as an introduction to collaborative engagement practises.
2. The Global Employee Fellowship programme (EFP), personally endorsed by Rio Tinto's CEO, has encouraged 140 employees of Rio Tinto's worldwide operations, since 1999, to participate in conservation projects that Earthwatch is funding around the world. The partnership "offers the opportunity to employees to make a practical contribution to a conservation project, while enhancing their skills and understanding of biodiversity, sustainability and corporate social and environmental responsibility" (Earthwatch, 2002a 43:1). The EFP has an impressive personal impact on the company's employees[2] ranging from increasing their interest in the environment and conservation issues to submitting their resignation to the company upon their return.
3. The African Fellowship Programme, whereby 13 African conservation professionals are supported by Rio Tinto to join Earthwatch projects in Africa, which are specifically designed for training purposes. Earthwatch's NGO partners in Africa select the Fellows. The initiative is part of a wider, well-established programme funded by the EU, the UK government and other corporate funders with the aim of building capacity in biodiversity management.
4. Field project grants. The company contributes financially to all biodiversity projects in which its own employees participate in order for project scientists to use the money according to their discretion.

---

[1] The components of the partnership presented below refer exclusively to the partnership between the Earthwatch Institute in Europe and Rio Tinto.

[2] Each employee completes a questionnaire upon returning from their placements. These comments are compiled annually by Earthwatch and are included in the report to Rio Tinto.

5. Annual lecture series. Rio Tinto provides funding to support a series of three lectures per year, delivered at the Royal Geographical Society by scientists receiving support from Rio Tinto through the partnership. Approximately 300 people attend each event. The aim of this lecture series is to increase public awareness of biodiversity.
6. Core funding. Rio Tinto further supports the science department at Earthwatch. Also Rio Tinto funds a part-time position within the Corporate Programmes Department that took over the overall coordination for the partnership.
7. Project Development. Rio Tinto further covers the costs for the development of new Earthwatch field projects specifically for the company, in the areas of mining or exploration, and involving Rio Tinto employees as volunteers.

The above are the main components of the partnership; however the partnership extends beyond the specific programmes as they interrelate and are interdependent with the organisation's existing programmes, structures and people, and therefore influence the organisation in a complex way.

Appendix 4

The Prince's Trust: Royal Bank of Scotland Group Partnership Content

The Partnership started in 2001 and was the result of conversations between the executives at the highest levels of the organizations.

The partnership was built on the existing 'transactional' relationships in place with Natwest and RBS.

The final partnership formation comprised a £3.4 million cash support package over 3 years and the provision of a £7 million loan facility.

The partnership consists of seven main components:

- **Route 14–25** is the initiative designed to change the way in which The Prince's Trust works. It is the cornerstone of the partnership and is core to the 3-year strategy of the Trust. In essence it is a programme of organisational development.

- **Employee Involvement** by staff at RBS and throughout their divisions allows The Prince's Trust to increase the volunteer resource at their disposal. A target of recruiting 550 volunteers over 3 years was set by the partnership. The opportunities offered by The Prince's Trust have been mapped against the RBS core competencies framework and hence contribute directly to staff development.

- **Business Awards** is the recognition event held across the UK for businesses started by young entrepreneurs through The Prince's Trust. It is sponsored and supported by RBS, Natwest and Ulster Bank and local bank staff support each event. The national event is attended by HRH The Prince of Wales and the CEO of RBSG.

- **Business Start-up** funding from RBS is in addition to the revolving loan facility and allows for the direct funding of new businesses started by young people with the aid of The Prince's Trust.

- **Coutts** is one of the Wealth Management divisions of the Group and a very prestigious brand. The Prince's Trust and Coutts have a unique relationship whereby they leverage the brand of both organisations to encourage donations to business start-ups through dinner and lunch events.

- **Sponsorship** of discrete events in partnership with other supporters, in particular sport, is a growing element of the partnership. Natwest sponsor The Prince's Trust cricket initiative in association with the English Cricket Board and the Professional Cricketers Association.

- **Seconded IT resource** from RBS to The Prince's Trust allows the development of critical systems within a shorter timeframe. The area of systems development is further supported by the relationships between RBS and their IT suppliers, giving RBS leverage used for their benefit.

- Further benefits from the partnership come through financial support and experience given during the development of the new The Prince's Trust brand. Also, the support of Senior RBS Executives in promoting The Prince's Trust work and introductions to public sector agencies – in both directions – and the funding of three members of the central project management team.

- It is also important to note that the RBS funding for Route 14–25 has enabled The Prince's Trust to leverage a significant sum of money from Public Sector sources such as Connexions, ESF and New Deal for Communities.

Bibliography

Abbott, A. (1992). From causes to events: Notes on narrative positivism. *Sociological Methods and Research, 20*, 428–455.

Adler, P. & Jermier, J. (2005). Developing a field with more soul: Standpoint theory and public policy research for management scholars. *Academy of Management Journal, 48*(6), 941–944.

Alsop, R. J. (2004). *The 18 immutable laws of corporate reputation.* New York: Free Press.

Alvarez, M., Binkley, E., Bivens, J., Highers, P., Poole, C., & Walker, P (1990). Case-based instruction and learning: An interdisciplinary project. *Proceedings of 34th Annual Conference* (pp. 2–18), College Reading Association. Reprint.

Alvesson, M. & Deetz, S. (2000). *Doing critical management research.* London: Sage.

Alvesson, M. & Stanely, D. (2000). *Doing critical management research.* Thousand Oaks, CA/London: Sage.

Andrioff, J. (2000). *Managing social risk through stakeholder partnership building.* Ph.D. thesis, Warwick University.

Anheir, H. (2000). *Managing nonprofit organisations: Towards a new approach* (Civil Society Working Paper 1). London: Centre for Civil Society, London School of Economics.

Anheir, H. (2001). *The voluntary sector class notes.* London: Centre for Civil Society, London School of Economics.

Argyris, C. (1985). *Strategy, change, and defensive routines.* Marshfield, MA: Pitman.

Astey, W. G. & Fombrun C. J. (1983). Collective strategy: Social ecology of organizational environments. *Academy of Management Review, 894*, 576–587.

Austin, J. E. (2000). Strategic collaboration between nonprofits and businesses. *Nonprofit and Voluntary Sector Quarterly, 29*(Suppl 1), 69–97.

Backer, L. (2007). Engaging Stakeholders in Corporate Environmental Governance. *Business and Society Review, 112*(1), 29–54.

Bank Track (2004). The equator principles. Bank Track website, from http://www.banktrack.org/doc/File/Policies%20and%20processes/Equator%20Principles/The%20Equator%20Principles.pdf. Accessed 19 Nov 2004

Barnett, W. P. & Carroll, G. R. (1995). Modelling internal organizational change. *Annual Review of Sociology, 21*, 217–236.

Barbier, E. (1987). The concept of sustainable economic development. *Environmental Conservation 14*(2), 101–110.

Barzelay, M., Gaetani, F., Cortazar-Velarde, J. C., & Cejudo, G. (2001). *Research on public management policy change in the Latin American region: Conceptual framework, methodological guide, and exemplars.* Retrieved 10, Jan 2002, from http://www.lse.ac.uk/collections/MES/pdf/IADBMethodGuide.pdf

Basler, D., & Carmin, J. (2002). The interpretive basis of action: Identity and tactics in environmental movement organizations. *Academy of Management Proceedings 2002.* Best Paper Index: B1–B6.

BBC News (2004). RBS unveils record £6.2bn profits. *BBC News, UK edition website.* Retrieved November 19, 2004, from http://www.bbc.co.uk

BBC News (2004). Royal Bank of Scotland has unveiled a forecast-beating jump in annual pre-tax profit to £6.2bn ($11.7bn). Retrieved February 19, 2004, from: http://66.102.9.132/search?q=cache: z_gJjkG6b0YJ:news.bbc.co.uk/2/hi/business/3502091.stm+bbc+news+2004+and+questioned+ whether+banks+and+credit+card+companies+were+taking+social+responsibility+seriously&c d=1&hl=en&ct=clnk&gl=uk

Bendell, J. (1998, July 8–11). *Citizens cane? Relations between business and civil society.* Paper presented at ISTR 3rd international conference, Geneva.

Bendell, J. (2000c). Working with stakeholder pressure for sustainable development. In J. Bendell (Ed.), *Terms of endearment. Business, NGOs and sustainable development.* Sheffield: Greenleaf Publishing.

Bendell, J. (2000a). A no win-win situation? GMOs, NGOs and sustainable development. In J. Bendell (Ed.), *Terms of endearment. Business, NGOs and sustainable. development.* Sheffield: Greenleaf Publishing.

Bendell, J. (2000b). Civil regulation: A new form of democratic governance for the global economy? In J. Bendell (Ed.), *Terms of endearment. Business, NGOs and sustainable development.* Sheffield: Greenleaf Publishing.

Bendell, J., & Lake, R. (2000). New frontiers: Emerging NGO activities and accountability in Business. In J. Bendell (Ed.), *Terms of endearment. Business, NGOs and sustainable development.* Sheffield: Greenleaf Publishing.

Bendell, J., & Murphy, D. F. (2000). Planting the seeds of change: Business-NGO relations on tropical deforestation. In J. Bendell (Ed.), *Terms of endearment. Business, NGOs and sustainable development* (pp. 65–78). Sheffield: Greenleaf Publishing.

Bennett, R. & Sargeant, A. (2003). The nonprofit marketing landscape: Guest editors' introduction to a special edition. *Journal of Business Research, 58,* 797–805.

Bennett, R. & Savani, S. (2004). Managing conflict between marketing and other functions within charitable organisations. *The Leadership & Organisation Development Journal, 25*(2), 180–200.

Berger, I. E., Cunningham, P. H., & Drumwright, M. E. (2004). Social alliances: Company/ nonprofit collaboration. *California Management Review, 47*(1), 58–90.

Berglind, M. & Nakata, C. (2005). Cause-related marketing. More buck than bang? *Business Horizons, 48,* 443–53.

Bevir, M. & Rhodes, R. A. W. (2006). Defending interpretation. *European Political Science, 5*(1), 69–83.

Biermann, F., Mol, A. P. J., & Glasbergen, P. (2007). Conclusion: Partnerships for sustainable development. In F. Biermann, A.P.J. Mol, & P. Glasbergen (Ed.), *Partnerships, governance and sustainable development.* Cheltenham: Edward Elgar.

Billis, D. & MacKeith, J. (1993). Organising NGOs: Challenges and trends in the management of overseas Aid. London, Centre for Voluntary Organisations, London School of Economics.

Birch, D. (2003). *Doing business in new ways. The theory and practice of strategic corporate citizenship with specific reference to Rio Tinto's community partnerships* (A Monograph). Melbourne: Corporate Citizenship Unit, Deakin University.

Black, J. (2002). *Critical reflections on regulation. Centre for analysis of risk and regulation (CARR)* (Discussion Paper 4). London: London School of Economics.

Blaza, A., Horrax S., & Hurt, H. (2002). *It's your choice! Influencing more sustainable patterns of production and consumption in the UK. A report on multistakeholder processes conducted by Imperial College and UNED-UK Committee.* Retrieved February 15, 2003, from http://www.env.ic.ac.uk/research/epmg/ItsYourChoice.pdf#search='purchasing%20 power:%20civil%20action%20for%20sustainable%20consumption'

Booth, S. A. (2000). How can organisations prepare for reputational crises? *Journal of Contingencies and Crisis Management, 8*(4), 197–207.

Bovaird, T., Loffler, E., & Parrado-Diez, S. (2002). Finding a bowling partner. The role of stakeholders in activating civil society in German, Spain and the UK. *Public Management Review, 4*(3), 411–431.

Brammer, S. & Millington, A. (2004). The development of corporate charitable contributions in the UK: A stakeholder analysis. *Journal of Management Studies, 41*, 1411–1434.

Bray, J. (2000). Web wars. NGOs, companies and governments in the internet-connected world. In J. Bendell (Ed.), *Terms of endearment. Business, NGOs and sustainable development.* Sheffield: Greenleaf Publishing.

Brown, S. L. & Eisenhardt, K. (1997). The art of continuous change: Linking complexity theory and time-paced evolution in relentlessly shifting organisations. *Administrative Science Quarterly, 42*, 1–34.

Brown, S. L. & Eisenhardt, K. (1998). *Competing on the edge.* Boston: Harvard Business School Press.

Bryman, A. (1988). *Quantity and quality in social research.* London/Boston: Unwin Hyman.

Bryman, A., & Bell, E. (2007). Business Research Methods (second edition). Oxford University Press.

Bryson, J. M., Gibbons, M. J., & Shaye, G. (2001). Enterprise schemes for nonprofit survival, growth and effectiveness. *Nonprofit Management and Leadership, 11*(3), 271–288.

Burrell, G. & Morgan, G. (1985). *Sociological paradigms and organizational analysis.* London: Heinemann.

Calhoun, C. (2002). *Politics. Dictionary of the social sciences.* Oxford University Press, Oxford Reference Online. Retrieved March 31, 2006, from http://www.oxfordreference.com/views/ENTRY.html?subview=Main&entry=t104.e1287

Cannon, T. (1992). *Corporate responsibility.* London: Pitman.

Caplan, K. (2003). *The purist partnership: Debunking the terminology of partnerships. In partnership matters. Current issues in cross-sector collaboration.* Copenhagen: Copenhagen Centre.

Caplan, K., & Jones, D. (2002). *Partnership indicators. Measuring effectiveness of multi-sector approaches to service provision. BPD Water and Sanitation Cluster. Practitioner note series: Partnership indicators.* London

Carriga, E. & Mele, D. (2004). Corporate social responsibility theories: Mapping the territory. *Journal of Business Ethics, 53*, 51–71.

Carroll, A. B. (1994). Social issues in management research. *Business and Society, 33*(1), 5–25.

Carmin, J. & Basler, D. (2002). Selecting repertoires of action in environmental movement organisations. *Organization and Environment, 15*(4), 365–388.

Charity Commission (2005, November). *Report of findings of a survey of public trust and confidence in charities.* Charity Commission.

Charles C. Ragin (1987). The comparative Method. Moving Beyond Qualitative and Quantitative Strategies. The Regents of the University of California.

Collins, D. (1996). New paradigms for change? Theories of organization and the organization of theories. *Journal of Organizational Change Management, 9*(4), 9–23.

Commission of the European Communities (2001). Green paper promoting a European framework for corporate social responsibility. Brussels, COM (2001) 416 final.

Corning, P. A. (2003, September 17–18). *The basic problem is still survival, and an evolutionary ethics is indispensable.* Paper presented at the Complexity, Ethics and Creativity conference. London: London School of Economics.

Coutoupis, T. (1996). *Choregia. A practical guide for sponsors and sponsored.* Athens: Galeos Publications.

Covey, J., & Brown, L. D. (2001). *Critical co-operation: An alterative form of civil society-business engagement* (IDR Reports, Vol. *17*(1)). London: Institute for Development Research.

Crane, A. (1998). Exploring green alliances. *Journal of Marketing Management, 14*(6), 559–579.

Crane, A. (2000). Culture clash and mediation. Exploring the culture dynamics of business-NGO collaboration. In J. Bendell (Ed.), *Terms of endearment. Business, NGOs and sustainable development.* Sheffield, UK: Greenleaf Publishing.

Crane, A., & Matten, D. (2004). *Business ethics. A European perspective. Managing corporate citizenship and sustainability in the age of globalization.* Oxford: Oxford University Press.

Creswell, J. (1994). *Research design: Qualitative and quantitative approaches*. Thousand Oaks, CA: Sage.

Dawson, P. (1994). *Organisational change: A processual approach*. London: Routledge.

Deakin, N. (2002). Public-private partnerships. A UK case study. *Public Management Review, 4*(2), 133–147.

Dechalert, P. (2002). *Managing for survival?: NGOs and organisational change. Case Studies of four small Thai NGOs*. Unpublished Ph.D. thesis, Centre for Civil Society, Social Policy Department, London School of Economics and Political Sciences.

Deephouse, D. L. & Carter, S. M. (2005). An examination of differences between organizational legitimacy and organizational reputation. *Journal of Management Studies, 42*(2), 329–360.

Dees, J. G. (1998). Enterprising nonprofits. *Harvard Business Review, 76*(1), 54–67.

Denzin, N., & Lincoln, Y. S. (Eds.). (2000). *Handbook of Qualitative Research* (2nd ed.). Thousand Oaks, CA: Sage.

Di Maggio, P. & Anheier, H. (1990). The sociology of the nonprofit sector. *Annual Review of Sociology, 16*, 137–159.

DiMaggio, P. & Powell, W. (1983). The iron cage revisited: Institutional isomorphism and collective rationality in organizational fields. *American Sociological Review, 48*, 147–160.

Doane, D. (2001). *Taking flight: The rapid growth of ethical consumption. The ethical purchasing index 2001. New economics foundation report for the co-operative bank*. Retrieved March 24, 2002, from http://www.neweconomics.org/gen/uploads/Taking%20Flight%20-%20EPI%202001.pdf

Dobson, A. (1998). *Justice and the environment: Conceptions of environmental sustainability and social justice*. Oxford: Oxford University Press.

Donaldson, T. & Preston, L. E. (1995). The stakeholder theory of the corporation: Concepts, evidence, and implications. *Academy of Management Review, 20*(1), 65–91.

Doh, J. P. & Teegen, H. (2002). Nongovernmental organizations as institutional actors in international business: Theory and implications. *International Business Review, 11*, 665–684.

Dreiling, M., & Wolf, B. (2001). Environmental movement organisations and political strategy. Tactical conflicts over NAFTA. *Organization & Environment, 14*(1), 34–54.

Drew, R. (2003). Learning in Partnership: What Constitutes learning in the Context of South-North Partnerships. A Discussion Paper: Submitted to BOND/the Exchange Programme; pp. 1–28.

Drucker, P. E. (1989). What business can learn from nonprofits. *Harvard Business Review*, July–August, 88–93.

Dryzek, J. S. (2005). *The politics of the earth. Environmental iscourses* (2nd ed.). Oxford: Oxford University Press.

Dutton, J. E. & Dukerich, J. M. (1991). Keeping an Eye on the Mirror: Image and identity in organizational Adaptation. *Academy of management Journal, 34*(3), 517–554.

Dutton, J. E., Ashford, S. J., O'Neill, R. M., & Lawrence, K. A. (2001). Moves that matter: Issue selling and organizational change. *Academy of Management Journal, 4*(44), 716–737.

Earthwatch (2001). *Earthwatch Annual Report 2001*. Oxford: Earthwatch Institute.

Earthwatch (2001R). *Report to Rio Tinto on the partnership between Rio Tinto and Earthwatch institute (Europe) between January 1999 and March 2001*. Oxford: Earthwatch.

Earthwatch (2002). *Implementing corporate environmental responsibility*. Directory of members of the Corporate Environmental Responsibility Group 2002–3003, Bulletin No.20. Oxford: Earthwatch.

Earthwatch (2003a). *Building bridges. Making connections to a sustainable future* (2003 Annual Report). Oxford: Earthwatch.

Earthwatch (2003Rb). *Report to Rio Tinto on 2002 programmes of activities*. Oxford: Earthwatch.

Earthwatch (2004a). *Corporate Programmes@Earthwatch Institute. Earthwatch website*. Retrieved July 3, 2004, from http://www.earthwatch.org/corporate/

Earthwatch (2004b). *The corporate environmental responsibility group. Earthwatch website*. Retrieved July 8, 2004, from http://www.earthwatch.org/europe/corporate/aboutcerg.html

Earthwatch (2004c). *Earthwatch's mission. Earthwatch website*. Retrieved July 3, 2004, from http://www.earthwatch.org/europe/aboutus.html

eFinancial Careers (2003). *Employer profile: Royal bank of Scotland. eFinancial Careers web site*. Retrieved November 19, 2004, from http://www.efinancialcareers.com/article_270.cfm?s toryref=17000000000016575&xsection=17&subsection=11

Ebrahim, E. (2005). *NGOs and organizational change. Discourse, reporting and learning*. Cambridge: Cambridge University Press.

Edwards, M. & Gaventa, J. (2001). *Global citizen action*. London: Earthscan.

Elkington, J. (1999). *Cannibals with forks: The triple bottom line of 21st century business*. London: Capstone.

Elkington, J., & Fennell, S. (2000). Seeking and managing collaboration. In J. Bendell (Ed.). *Terms of endearment. Business, NGOs and sustainable development*. Sheffield: Greenleaf Publishing.

Eurobarometer 61 (2004). *Public opinion in the European union*. Field Work: February–March 2004. Publication July 2004.

Evered, R., & Louis, M. R. (1991). Alternative perspective in the organizational sciences: 'Inquiry from the inside' and 'inquiry from the outside'. In N.C. Smith, & P. Dainty (Eds.), *The management research handbook*. London/New York: Routledge.

Fabig, H. & Boele, R. (1999). The changing nature of NGO activity in a globalizing world: pushing the corporate responsibility agenda. *IDS Bulletin, 30*, 58–67.

Farrar, J. H. & Hannigan, B. M. (1998). *Farrar's company law* (4th ed.). London: Butterworths.

Fielding, N. G. & Lee, R. M. (1998). *Computer analysis and qualitative research*. London: Sage.

Fishel, D. (1993). *The arts sponsorship handbook*. London: Directory of Social Change.

Fisher, R. & Ury, W. (1981). *Getting to yes: Negotiating agreement without giving in*. Boston: Houghton Mifflin.

Flagan, M. (2004). RBS faces criticism as profits hit £7bn. *News. Scotsman.com website*. Retrieved November 19, 2004, from:http://www.scotsman.com

Flick, U. (1998). *An introduction to qualitative research*. London: Sage.

FoE (1995). *Press release: FOE's RTZ Shareholder Campaign*. FoE website. Retrieved July 6, 2004, from http://www.foe.co.uk/resource/press_releases/19950608145818.html

FoE (2004). *Press release: UK banks failing to implement green equator principles*. FoE website. Retrieved November 19, 2004, from http://www.foe.co.uk/resource/press_releases/uk_banks_ failing_to_implem_03062004.html

Fowler, A. (2000). Beyond partnership: Getting real about NGO relationship in the aid system. *IDS Bulletin, 31*(3).

Fowler, P., & Heap, S. (2000). Bridging troubled waters: The Marine Stewardship council. In J. Bendell (Ed.), *Terms of endearment. Business, NGOs and sustainable development*. Sheffield: Greenleaf Publishing.

Freeman, R. E. (1984). *Strategic management: A stakeholder approach*. Boston: Pitman.

Friedman, M. (1970). A Friedman's doctrine – The social responsibility of business is to increase its profits. *New York Times* 32–33, 122–126.

FTSE4GOOD (2003). *FTSE4GOOD index series. Inclusion criteria*. London: FTSE The Index Company. Retrieved July 2006, from http://www.ftse.com/Indices/FTSE4Good_Index_Series/ Downloads/FTSE4Good_Inclusion_Criteria_Brochure_Feb_03.pdf

Galaskiewicz, J., & Sinclair Colman, M. (2006). Collaboration between corporations and nonprofit organizations. In R. Steinberg, & W.W Powel (Eds.), *The nonprofit sector. A research handbook* (2nd ed.). New Haven, CT: Yale University Press.

Galaskiewicz, J. & Wasserman, S. (1989). Mimetic processes within an interorganizational field: An empirical test. *Administrative Science Quarterly, 34*, 454–479.

Galbreath, J. R. (2002). Twenty-first century management rules: The management of relationships as intangible assets. *Management Decision, 40*(2), 116–126.

Galbreath, J. R. (1977). *Organization design*. Reading, MA: Addison-Wesley.

Garney, G. (1998). *Legislators, ministers and public officials*. (Working Paper Series, Transparency International). Retrieved March 25, 2001, from http://www.transparency.org/working_papers/ carney/index.html

Gephart, R. P. (1999). Paradigms and research methods. *Research methods forum, Vol. 4. On line issue*. Research methods division, Academy of Management. Retrieved February 30, 2001, from http://division.aomonline.org/rm/

Gibbs, G. R. (2002). *Qualitative data analysis. Explorations with NVivo*. Buckingham: Open University Press.

Giddens, A. (2001, Nov 7). *Director's lectures at the LSE*. Retrieved February, 2005, from http://old.lse.ac.uk/collections/meetthedirector/pdf/07-Nov-01.pdf

Giving List (2004). *Giving list 2004. Society Guardian website*. Retrieved November 20, 2004, from http://society.guardian.co.uk/givinglist/0,10994,579376,00.html

Gladwin, T. N. & Kennelly, J. J. (1995). Shifting paradigms for sustainable development: Implications for management theory and research. *Academy of Management Review, 20*(4), 874–904.

Glasbergen, P. (2007). Setting the scene: The partnership paradigm in the making. In P. Glasbergen, F. Birmann, & A. P. J. Mol, *Partnerships, governance and sustainable development 2007*. Cheltenham: Edward Elgar.

Googins, B. & Rochlin, S. (2000). Creating the partnership society: Understanding the rhetoric and reality of cross sector partnerships. *Business and Society Review, 105*(1), 127–144.

Gray, B. (1989). *Collaborating*. San Francisco: Jossey-Bass.

Greenall, D. & Rovere, D. (1999). *Engaging stakeholders and business-NGO partnerships in developing countries*. Ontario: Centre for innovation in Corporate Social Responsibility.

Greenwood, R. & Hinnings, C. R. (1996). Understanding radical change: Bringing together the old and the new institutionalism. *Academy of Management Review, 21*(4), 1022–1054.

Guba, E. G. & Lincoln, Y. S. (1994). Competing paradigms in qualitative research. In N. K. Denzin & Y. S. Lincoln (Eds.), *Handbook of qualitative research* (pp. 105–117). Newbury Park, CA: Sage.

Habermas, J. (1986). The theory of communicative action. *Reason and the Rationalisation of Society, 1*. Cambridge: Polity Press.

Hall, R. H., Clark, J. C., Giordano, P. C., Johnson, P. V., & Van Roekel, M. (1977). Patterns of interorganisational relationships. *Administrative Science Quarterly, 22*, 457–470.

Hamman, R. & Acutt, N. (2003). How should Civil Society (and the government) Respond to 'corporate social responsibility'? A critique of business motivations and the potential for partnerships. *Development Southern Africa, 20*(2), 255–270.

Hardis, J. (2003). Social multipartite partnerships. When practice does not fit rhetoric. In M. Morsing, & C. Thyssen (Eds.), *Corporate values and responsibility. The case of Denmark*. Copenhagen: Samfundslitteratur.

Hardy, C., Phillips, N., & Lawrence, T. (1998). Distinguishing trust and power in interorganizational relations: Forms and facades of trust. In C. Lane & R. Bachmann (Eds.), *Trust within and between Organizations* (pp. 64–87). Oxford: Oxford University Press.

Hart, J. (2000, March 20). Excess bank profits slammed. *Evening Standard website*. Retrieved November 19, 2004, from http://www.eveningstandard.co.uk

Hart, J. (2001). Protest over RBS bonus plans. *Evening Standard website*, 20 March 2000. Retrieved November 19, 2004, from http://www.eveningstandard.co.uk

Hartman, C. L. & Stafford, E. R. (1997). Green alliances: Building new business with environmental groups. *Long Range Planning, 30*(2), 184–196.

Hartman, C. L., Hofman, P. S., & Stafford, E. R. (1999). Partnership: Path to sustainability. *Business Strategy and the Environment, 8*, 255–266.

Hassard, J. S. (1991). Multiple paradigm analysis: A methodology for management research. In N. C. Smith & P. Dainty (Eds.), *The management research handbook*. London/New York: Routledge.

Hatch, M.J. (1997). *Organization theory. Modern, symbolic, and postmodern perspectives*. Oxford: Oxford University Press.

Hatch, M. J. & Schultz, M. (1997). Relations between organisational culture, identity and image. *European Journal of Marketing, 31*(5–6), 356–365.

Heap, S. (1998). *NGOs and the private sector: Potential for partnerships?* (INTRAC Occasional Papers Series No. 27). Oxford: Intrac Publications.

Heap, S. (2000). *NGOs engaging with business: A world of difference and a difference to the world*. Oxford: Intrac Publications.

Heracleous, L. & Barrett, M. (2001). Organizational change as discourse: Communicative actions and deep structures in the context of information technology implementation. *Academy of Management Journal, 44*, 755–778.

Hemphill, T. A. & Vonortas, N. S. (2003). Strategic research partnerships: A managerial perspective. *Technology Analysis and Strategic Management, 15*(2), 255–271.

Henn, M., Weinstein, M., & Wring, D. (2000). *A generation apart? Youth and political participation in Britain today*. Paper for the Political Studies Association-UK 50th annual conference 10–13 April 2000, London. Retrieved November 20, 2001, from http://www.psa.ac.uk/cps/2000/Henn%20Matt%20et%20al.pdf#search='crisis%20of%20legitimacy%20in%20britain

Himmelstein, J. L. (1997). *Looking good and doing good: Corporate philanthropy and corporate power*. Bloomington: Indiana University Press.

Holzer, B. (2001). *Translational subpolitics and corporate discourse: A study of environmental protest and the Royal Dutch/Shell Group*. Unpublished doctoral dissertation, Department of Sociology, London School of Economics, London.

Homan, R. (1991). *The ethics of social research*. London: Longman.

Howell, J. & Pearce, J. (2001). *Civil Society and development. A critical exploration*. Boulder, CO/London: Lynne Rienner.

Hunter, A. (1974). *Symbolic communities. The persistence and change in Chicago's local communities*. Chicago: University of Chicago Press.

Huxham, C. (1993). Collaborative capability: An intra-organizational perspective on collaborative advantage. *Public Money & Management, 13*(3), 21–28.

Huxham, C. & Vangen, S. (1996). Working together. Key themes in the management of relationships between public and nonprofit organizations. *International Journal of Public Sector Management, 9*(7), 5–17.

Itami, H. & Roehl, T. (1987). *Mobilizing invisible assets*. Cambridge, MA: Harvard University Press.

Iyer, E. (2003). Theory of alliances: Partnerships and partner characteristics. *Journal of Nonprofit and Public Sector Marketing, 11*(1), 41–57.

Jenkins, R. (1996). *Social Identity*. London: Routledge.

Jupp, B. (2000). *Working together. Creating a better environment for cross-sector partnerships. A demos paper.*. London: Demos.

Kanter, R. M. (1994). Collaborative Advantage. Successful partnerships manage the relationship, not just the deal. *Harvard Business Review, July–August*, 96–108.

Kanter, R.M. (1999). From spare change to real change – the social sector as beta site for business innovation. *Harvard Business Review, May–June*, 122–132.

Katz, D. & Khan, R. L. (1978). *The social psychology of organisation* (2nd ed.). New York: Wiley.

Klein, N. (2000). *No logo*. New York: Picador Reading Books Guide.

Kolk, A., Van Tulder, R., & Kostwinder, E. (2008). Partnerships for development. *European Management Journal, 26*(4), 262–273.

Kotten, J. (1997). The strategic uses of corporate philanthropy. In L. Caywood Clark (Ed.), *The handbook of strategic public relations and integrated communications* (pp. 149–172). New York: McGraw-Hill.

Koza, M.P., & Lewin, A.Y. (1998). The Co-evolution of strategic alliances. *Organization Science, 9*(3), 255–263.

Krasner, S. (1995). Power politics, institutions and translational relations. In T. Risse-Kappen (Ed.), *Non-state actors, domestic structures and international institutions*. Ithaca, NY/London: Cambridge University Press.

Lamont, M. (1992). *Money. Morals and manners: The culture of the French and American upper-middle class*. Chicago: University of Chicago Press.

Lamont, M. & Molnar, V. (2002). The study of boundaries in the social sciences. *Annual Review of Sociology, 28*, 167–95.

Leiter, J. (2005). Structural isomorphism in Australian nonprofit organizations. voluntas. *International Journal of Voluntary and Nonprofit Organizations, 16*(1), 1–31.

Levine, S. & White, P. (1961). *Exchange as a conceptual framework for the study of interorganisational relationships*. New York: Holt, Rinehart &and Winston.

Lewin, A. Y., Long, C. P., & Carroll, T. N. (1999). The co-evolution of new organisational forms. *Organization Science, 10*(5), 535–550.

Lewis, D. (1999a). Introduction: The parallel universes of third sector research and the changing context of voluntary action. In D. Lewis (Ed.), *International perspectives on voluntary action: Reshaping the third sector*. London: Earthscan.

Lewis, D. (1999b). *International perspectives on voluntary action*. London: Earthscan.

Lewis, D. & Wallace, T. (2000). New Roles and Relevance. Development NGOs and the Challenge of Change. Kumarian Press.

Ling, T. (2002). Delivering joined-up government in the UK: Dimensions, issues and problems. *Public Administration, 80*(4), 615–642.

Lister, S. (2000). Power in partnership? An analysis of an NGO's relationships with its partners. *Journal of International Development, 12*, 227–239.

Lister, S. (2001). Consultation as a legitimising practice: A study of British NGOs in Guatemala. Ph.D. thesis, Centre for Civil Society, London School of Economics, London.

Long, F. J. & Arnold, M. B. (1995). *The power of environmental partnership*. Fort Worth: The Dryden Press.

Lovelace, K., Shapiro, D. L., & Weingart, L. R. (2001). Maximizing cross-functional new product teams' innovativeness and constraint adherence: A conflict communications perspective. *Academy of Management Journal, 44*(4), 779–793.

Loza, J. (2004). Business-community partnerships: The case for community organization capacity building. *Journal of Business Ethics, 53*, 207–311.

Macdonald, S. & Piekkari, R. (2005). Out of control: Personal networks in European collaboration. *R&D Management, 35*(4), 441–453.

MacKeith, J. (1993). *NGO Management, A guide through the literature*. London: Centre for Voluntary Organisations (CVO), London School of Economics and Political Sciences.

McWilliams, A. & Siegel, S. (2002). Corporate social responsibility: A theory of the firm Perspective. *Academy of Management Review, 26*, 117–127.

Mancuso Brehm, V. (2001). Promoting effective North-South NGO partnerships: A comparative study of 10 European NGOs (INTRAC The International NGO Training and Research Centre Occasional Papers Series No. 35), 1–75.

Marshall, G. (1998). *Dictionary of sociology*. Oxford: Oxford University Press.

Marshall, C. & Rossman, G. B. (1986). *Designing qualitative research*. Newbury Park, CA: Sage.

Martin, R. (2000). Breaking the code of change. Observations and critique. In M. Beer & N. Nohria (Eds.), *Breaking the code of change*. Boston: Harvard Business School Press.

Matten, D. & Crane, A. (2005). Corporate citizenship-towards an extended theoretical conceptualisation. *Academy of Management Review, 30*(1), 166–179.

Mazur, R. E. (2004). *Dictionary of critical sociology*. Retrieved February 15, 2004, from, http://www.public.iastate.edu/~rmazur/dictionary/a.html

McFarlan, F. W. (1999). *Working on nonprofit boards: Don't assume the shoe fits* (pp. 65–80). November–December: Harvard Business Review.

Mead, G. H. (1934). *Mind, self and society*. Chicago: University of Chicago Press.

Mead, G.H. (1936). Movements of thought in the nineteenth Century. In M.H. Moore (Ed.), *Introduction*. Chicago: University of Chicago Press.

Mead, G.H. (1938). *The philosophy of the act*, ed. C.W. Morris et al. Chicago: University of Chicago Press.

Meadowcroft, J. (2007). Democracy and accountability: The challenge for cross sector partnerships. In P. Glasbergen, F. Biermann, & A. P. J. Mol (Eds.), *Partnerships, governance and sustainable development*. Cheltenham: Edward Elgar.

Meadows, D. H., Meadows, D. L., & Randers, J. (1992). *Beyond the limits, envisioning a sustainable future*. Post Mills, VT: Chelsea Green.

Meenaghan, T. (1983). Commercial sponsorship. *European Journal of Marketing, 17*(7), 1–74.

Meenaghan, T. (1984). *Commercial sponsorship. Management bibliographies and reviews.* London: MCB Press.

Meyer, J., Boli, J., Thomas, G., & Ramirez, F. O. (1997). World society and the nation state. *American Journal of Sociology, 103*(1), 144–181.

Miles, M. & Huberman, A. (1994). *Qualitative data analysis: An expanded sourcebook.* London: Sage.

Millar, C., Choi, J. C., & Chen, S. (2004). Global strategic partnerships between MNEs and NGOs: Drivers of change and ethical issues. *Business and Society Review, 109*(4), 95–414.

Miller, C. & Ahmad, Y. (2000). Collaboration and partnership: An effective response to complexity and fragmentation or solution built on sand? *International Journal of Sociology and Social Policy, 20*(5/6), 1–38.

Miller, R. S. & Stephen, W. J. D. (1966). Spatial relationships in flocks of Sandhill Cranes (*Grus canadensis*). *Ecology, 47*(2), 323–327.

Mills, C. W. (1940). Situated actions and vocabularies of motive. *American Sociological Review, 5*(6), 904–913.

Milne, G. R., Iyer, E. S., & Gooding-Williams, S. (1996). Environmental organization alliance relationships within and across nonprofit, business and government sectors. *Journal of Public Policy and Marketing, 15*(2), 203–215.

Mir, R., Kashyap, R., & Iyer, E. (2006). Toward a responsive pedagogy: Linking social responsibility to firm performance issues in the classroom. *Academy of Management Learning & Education, 5*(3), 366–376.

Mitchell, K. & Arie, L. (2000). Managing partnerships and strategic alliances: raising the odds of success. *European Management Journal, 18*(2), 146–151.

Mitchell, R. K., Agle, B. R., & Wood, D. J. (1997). Toward a theory of stakeholder identification and salience: Defining the principle of who and what really counts. *Academy of Management Review, 22*(4), 853–886.

Mohiddin, A. (1998). Partnership: A new buzz word or realistic partnership? *Journal of the Society for International Development, 41*(4), 5–16.

Mol, A. P. J. (2007). Bringing the environmental state back in: Partnerships in perspective. In P. Glasbergen, F. Biermann, & A. P. J. Mol (Eds.), *Partnerships, governance and sustainable development.* Cheltenham: Edward Elgar.

Moody, R. (1991). *Plunder.* London: Partizans/CAFCA.

Moon, J. (2002). Business social responsibility and new governance. *Government and Opposition, 37*(3), 385–408.

Moon, J. (2004). Government as a driver for corporate social responsibility (Research Paper Series No. 22–2004). International centre for corporate social responsibility, Nottingham University Business School, University of Nottingham.

Moore, G. (1995). Corporate community involvement in the UK: Investment or atonement? *Business Ethics: A European Review, 4*(3), 171–178.

Morsing, M., & Thyssen, C. (Eds.) (2003). *Corporate values and responsibility. The case of Denmark.* Copenhagen: Samfundslitteratur.

Mortished, C. (2003). Miner may have dug itself into a hole. *The Times,* 23.

Moser, T. (2001). MNCs and sustainable business practice: The case of the Colombian and Peruvian petroleum industries. *World Development, 29*(2), 291–309.

Mowjee, T. (2001). *NGO-donor funding relationships: UK government and European community funding for the humanitarian aid activities of the UK NGOs from 1990–1997.* Social Policy Department, Centre for Civil society. London: London School of Economics.

Murphy D. F., & Bendell, J. (1999, Aug). *Partners in time? Business, NGOs and sustainable development. United Nations Research Institute for Social Development (UNRISD)* (Discussion Paper No. 109).

Murphy, D. F. & Bendell, J. (1997). *In the company of partners.* Bristol: The Policy Press.

Murphy, D. F., & Coleman, G. (2000). Thinking partners: Business, NGOs and the partnership concept. In J. Bendell (Ed.), *Terms of endearment. business, NGOs and sustainable development*. Sheffield: Greenleaf Publishing.

Nelson, J. & Zadek, S. (2000). *Partnership Alchemy: New social partnerships in Europe*. Copenhagen: The Copenhagen Centre.

Newell, P. (2000). Globalisation and the new politics of sustainable development. In J. Bendell (Ed.), *Terms of endearment. Business, NGOs and sustainable development*. Sheffield: Greenleaf Publishing.

Oliver, C. (1990). Determinants of interorganisational relationships: Integration and future directions. *Academy of Management Review, 15*(2), 241–265.

Oliver, P. E. & Johnston, H. (2000). What a good idea: Frames and ideologies in social movement research. *Mobilization: An International Journal, 5*, 37–54.

Oster, S. M. (1995). *Strategic management for nonprofit organizations*. New York: Oxford University Press.

Parker, B. & Selsky, J. (2004). Interface dynamics in cause based partnerships: An exploration of emergent culture. *NonProfit and Voluntary Sector Quarterly, 33*, 458–486.

Partizans (2003). *About us*. Retrieved April 3, 2005, from, http://www.minesandcommunities.org/Aboutus/partisans.htm

Patterson, L. K. (2004). Organizational effectiveness and leadership roles in two-year colleges in Tennessee. University of Memphis, Ed. D., 2003. Dissertation Abstracts International DAI 64 (10), 3576A. (Accession No. AAI3108561)

Patton, M. Q. (1980). *Qualitative evaluation and research methods*. London: Sage.

Perrow, C. (1994). Dialogue: Pfeffer slips! *Academy of Management Review, 19*, 191–204.

Peters, F. E. (1967). *Greek philosophical terms. A historical lexicon*. New York: New York University Press.

Peters-Okleshen, C. L. (2004). Using vocabularies of motives to facilitate relationship marketing: The context of the Winnebago Itasca travellers club. *Journal of Vocation Marketing, 10*(3), 209–222.

Pettigrew, A. (1985). Contextualist research and the study of organisational change process. In E. Mumford, G. Fitzgerald, & T. Wood-Harper (Eds.), *Research methods in information systems*. Proceedings of the IFIP WG8.2 Colloquium, Manchester, UK, 1–3 September 1984. Elsevier Science Publishers B.V.

Pettigrew, A. (1990). Is corporate culture manageable? In R. H. Rosenfeld, & D. C. Wilson (Eds.), *Managing organisations*. London: McGraw-Hill.

Pettigrew, A. M. (2000). Linking change processes to outcomes. A commentary on Ghoshal, Barlett, and WeickIn. In M. Beer, & N. Nohria (Eds.) *Breaking the code of change* (pp. 243–265). Boston: Harvard Business School Press.

Pettigrew, A. M., Woodman, R. W., & Cameron, K. S. (2001). Studying organisational change and development: Challenges for future research. *Academy of Management Review, 44*(4), 697–713.

Pinder, C., & Bourgeoise, V. (1982). Borrowing and the effectiveness of administrative science (Working Paper No. 848). New York: University of British Columbia.

Pondy, L. R. (1967). Organizational conflict. *Concepts and Models. Administrative Science Quarterly, 12*(2), 296–320.

Pondy, L. R. (1969). Varieties of organizational conflict. *Administrative Science Quarterly, 14*(4), 499–506.

Poole, M. S., Van de Ven, A. H., Dooley, K., & Holmes, M. H. (2000). *Organisation change and innovation processes*. Oxford: Oxford University Press.

Post, J. E., & Griffin, J. J. (1997). Corporate reputation and external affairs management. *Corporate Reputation Review, 1*(1), 165–71.

Powell, W. W. & Clemens, S. (1998). *Private action and the public good*. New Haven, CT: Yale University Press.

Power, M. (2004). *The risk management of everything. A Demos Publication*. London: Demos.

Prahalad, C. K., & Hamel, G. (1990). The core competence of the corporation. *Harvard Business Review*, May–June, pp. 71–91.

Pratt, B. & Loizos, P. (1992). *Choosing research methods: Data collection for development workers*. Oxford: Oxfam Print Unit.

The Prince's Trust (2003). *Breaking barriers? Reaching the hardest to reach* (Research prepared and printed by The Prince's Trust). London: The Prince's Trust.

The Prince's Trust. (2003a). *Our brand*. London: The Prince's Trust.

The Prince's Trust, 2004a. Our president: The Prince of Wales. The Prince's Trust website. Available from: http://www.princestrust.org.uk/Main%20Site%20v2/about%20us/the%20 prince%20of%20wales.asp. Accessed 15 November 2004.

The Prince's Trust. (2004). *2004/5 Fact sheet*. London: The Prince's Trust.

The Prince's Trust (2004c). *Trust, RBS announce £5 million enterprise initiative*. The Prince's Trust website. Retrieved November 22, 2004, from http://www.princestrust.org.uk/Main%20 Site%20v2/headline%20news/media%20room.asp.

The Prince's Trust (2004d). *The Prince's Trust Annual Review 2004*. The Prince's Trust website. Retrieved January 20, 2004, from: http://83.138.128.16/Review/Annual%20Review%2004.swf

Punch, M. (1994). Politics and ethics in qualitative research. In N. Denzin, & Y. S. Lincoln (Eds.), *Handbook of qualitative research* (2nd ed.). Thousand Oaks, CA/London: Sage.

Ragin, C. C. (1991). *Issues and alternatives in comparative social research*. Leiden, The Netherlands/ New York: E. J. Brill.

Ragin, C. C. & Becker, H. S. (1995). *What is a case? Exploring the foundations of social inquiry* (3rd ed.). Cambridge: Cambridge University Press.

RBSG (2004a). A brief history of the Royal Bank of Scotland Group. Royal Bank of Scotland website. Retrieved November 17, 2004, from http://www.rbs.co.uk/Group_Information/ Memory_bank/Our_history

RBSG. (2004b). *Corporate responsibility report 2003*. London: Royal Bank of Scotland publication.

RBSG. (2004c). *Annual report and accounts 2003*. London: Royal Bank of Scotland publication.

Rees, S. (1991). *Achieving power: Practice and policy in social welfare*. North Sidney, Australia: Allen & Unwin.

Rio, T. (2002). Social and environmental review highlights. Rio tinto publication.

Risse, T. (2001). Transnational actors, networks, and global governance. In W. Carlsnaes, T. Risse & B. Simmons (Eds.), *Handbook of international relations*. London: Sage.

Robbins, S. P. (2005). *Organizational behavior* (11th ed.). Englewood Cliffs, NJ: Prentice-Hall.

Rondinelli, D. A. & London, T. (2003). How corporations and environmental groups cooperate: Assessing cross-sector alliances and collaborations. *Academy of Management Executive, 17*(1), 61–76.

Rowe, M. & Devanney, C. (2003). Partnership and the governance of regeneration. *Critical Social Policy, 23*(3), 375–397.

Rowley, T. J. (1997). Moving beyond dyadic ties: A network theory of stakeholder influences. *Academy of Management Review, 22*(4), 887–911.

Solomon, J. (2004). *Corporate governance and accountability* (2nd ed.). Chichester: Willey.

Salamon, L. M. (1992). *America's nonprofit sector: A primer*. New York: Foundation Center.

Salamon, L. M. & Anheier, H. K. (1997). *Defining the nonprofit sector: A cross-national analysis*. Manchester: Manchester University Press.

Salamon, L. M., Anheier, H. K., List, R., Toepler, S. W., et al. (1999). *Global civil society: Dimension of the nonprofit sector*. Baltimore, Maryland: Institute of Policy Studies, John Hopkins University.

Sanchez, P., Chaminade, C., & Olea, M. (2000). Management of intangibles. An attempt to build a theory. *Journal of International Capital, 1*(4), 312–327.

Sargeant, A. (1995). Do UK Charities have a lot to learn? *Fund Raising Management, 26*(5), 14–16.

Sartori, G. (1994). Compare why and how: Comparing, miscomparing and the comparative method. In M. Dogan & A. Kazancigil (Eds.), *Comparing nations: Concepts, strategies, substance*. Oxford: Blackwell.

Sastry, M. A. (1997). Problems and paradoxes in a model of punctuated organizational changes. *Administrative Science Quarterly, 42*, 237–275.

Schneidewind, U., & Petersen, H. (2000). Change the rules! Business-NGO partnerships and structuration theory. In J. Bendell (Ed.), *Terms of endearment. Business, NGOs and sustainable development.* Sheffield: Greenleaf Publishing.

Schön, D. A. (1991). *The reflective turn: Case studies in and on educational practice.* New York: Teachers Press, Columbia University.

Schuler, D. A. & Cording, M. (2006). A corporate social performance-corporate financial performance behavioral model for consumers. *Academy of Management Review, 31*(3), 540–558.

Schwartz, H. & Davis, S. M. (1981). Matching corporate culture and business strategy. *Organizational Dynamics, 10*(1), 30–48.

Scott, J., & Marshall, G. (2005). *Political science. A dictionary of sociology.* Oxford Reference online, Oxford University Press. Retrieved March 31, 2006, from http://www.oxfordreference.com/views/ENTRY.html?subview=Main&Entry=t88.e1742

Seitanidi, M. M. (2008). Adaptive responsibilities: Non-linear interactions across social sectors. Cases from cross sector partnerships. *Emergence: Complexity & Organization (E: CO) Journal, 10*(3).

Seitanidi, M. M. (2009). Employee involvement in implementing CSR in cross sector social partnerships. *Corporate Reputation Review* (in press).

Seitanidi, M. M., & Crane, A. (2009). Corporate social responsibility in action. Partnership management: Selection-design-institutionalisation. *Journal of Business Ethics, Special Issue on: CSR Implementation,* (DOI 10.1007/s10551-008-9743-y).

Seitanidi, M. M. & Ryan, A. M. (2007). A critical review of forms of corporate community involvement: From philanthropy to partnerships. *International Journal of Nonprofit and Voluntary Sector Marketing, 12*(3), 247–266.

Seitanidi, M. M. (2007a). Intangible economy: How can investors deliver change in businesses? Lessons from nonprofit-business partnerships. *Management Decision, 45*(5), 853–865.

Seitanidi, M. M. (2007b). *The future challenges of cross sector interactions: Interactions between nonprofit organisations and businesses* (BRESE Working Paper Series No. 22). Brunel: Brunel Business School, Brunel University.

Seitanidi, M. M. (2005). Corporate social responsibility and the non-commercial sector. What does corporate social responsibility mean for the non-commercial sector? And is it different from the CSR for businesses? *New Academy Review, 3*(4), 60–72.

Selsky, J. W. & Parker, B. (2005). Cross-sector partnerships to address social issues: Challenges to theory and practice. *Journal of Management, 31*(6), 1–25.

Senior, B. & Fleming, J. (2006). *Organizational change* (3rd ed.). Englewood Cliffs, NJ: Prentice-Hall.

Shaffer, B. & Hillman, A. J. (2000). The development of business-government strategies by diversified firms. *Strategic Management Journal, 21*(2), 175–19.

Shaw, R. (1993). *The spread of sponsorship.* Northumberland, UK: Bloodaxe Books.

Shen, J. (2005). Expanding the frontier of global knowledge: Introduction. *Journal of Organisational Transformation and Social Change, 2*(1), 3–8.

Silverman, D. (1970). *The theory of organizations.* London: Heinemann.

Silverman, D. (1993). *Interpreting qualitative data. Methods for analysing talk, text and interaction.* London: Sage.

Spruyt, H. (1994). *The sovereign state and its competitors.* Princeton, NJ: Princeton University Press.

Stafford, E. R. & Hartman, C. L. (1996). Green alliances: Strategic relations between business and environmental groups. *Business Horizons, 39*(2), 50–59.

Stafford, E. R., & Hartman, C. L. (1998) Toward an understanding of the antecedents of environmentalist-business cooperative relations. In R. C. Goodstein, & S. B. MacKenzie (Eds.), *American marketing association summer educators' conference Proceedings* (pp. 56–63). Chicago: American Marketing Association.

Stafford, E. R., & Hartman, C. L. (2000). Environmentalist-business collaborations: Social responsibility, green alliances, and beyond. In G. Zinkhan (Ed.), *Advertising research: The Internet, consumer behavior, and strategy* (pp. 170–192). Chicago: American Marketing Association.

Stafford, E. R., & Hartman, C. L. (2001). Greenpeace's 'Greenfreeze Campaign'. Hurdling competitive forces in the diffusion of environmental technology innovation. In K. Green, P. Groenewegen, & P. Hofman (Eds.), *Ahead of the curve*. AH Dordrecht, The Netherlands: Kluwer.

Stacey, R. D. (2001). *Complex responsive processes in organizations. Learning and knowledge creation*. London/New York: Routledge.

Strategy Unit (2002). *Private action, Public benefit: A review of charities and the wider not-for-profit sector. Prime minister's strategy unit*. Cabinet Office. Retrieved September 2004, from http://www.number-10.gov.uk/su/voluntary/report/downloads/strat-data.pdf

Sullivan, H. & Skelcher, C. (2002). Working across boundaries, collaboration in public services. Palgrave Macmillan.

Sussex, J. (2003). Public-private partnerships in hospital development: Lessons from the UK's 'Private Finance Initiative'. *Research in Healthcare Financial Management, 8*(1), 59–76.

Tarrow, S. (2001). Translational politics: Contention and institutions in international politics. *Annual Review of Political Science, 4*, 1–20.

Third Sector (2001). *Interview*. Appeared on the issue of 21 January 2001. Available from: http://www.thirdsector.co.uk

Third Sector (2003). *WWF rejects funds in quarry row*. Appeared 3 December 2003. Retrieved December 4, 2005, from http://www.thirdsector.co.uk/charity_news/full_news.cfm?ID=8781

Third Sector (2005a). *Charities fret over corporate tie-ups*. 9 March 2005. Retrieved April 10, 2006, from http://www.thirdsector.co.uk/charity_news/full_news.cfm?ID=13844

Third Sector (2005b). *Corporate responsibility-patience is the essential ingredient in partnerships*. Appeared 2nd November 2005. Retrieved November 25, 2005, from http://www.thirdsector.co.uk/charity_news/full_news.cfm?ID=16692

Torchim, W. M. (2002). *The research method knowledge base* (2nd ed.) Retrieved August 5, 2002, from URL: <http://trochim.human.cornell.edu/kb/positivism.htm

Tosi, H. L., Rizzo, J. R., & Carroll, S. J. (1994). *Managing organizational behavior*. Oxford: Blackwell.

Trevino, L. K. & Nelson, K. A. (2007). *Managing business ethics. Straight talk about how to do it right* (4th ed.). Hoboken, NJ: Wiley.

Tully, S. (2004). *Corporate-NGO partnerships as a form of civil regulation: Lessons from the energy and biodiversity initiative* (Discussion Paper No. 22), June 2004. London: ESRC Centre for Analysis of Risk and Regulation (CARR), London School of Economics.

Turcotte, M. F. (2000). Working non-'Stop' for sustainable development: Case study of the Canadian environmental NGO's relationship with business since 1970. In J. Bendell (Ed.), *Terms of endearment. Business, NGOs and sustainable development*. Sheffield: Greenleaf Publishing.

Tyson, S. & Jackson, T. (1992). *The essence of organisational behaviour*. Englewood Cliffs, NJ: Prentice-Hall.

Uri, J. (1991). Time and space in Gidden's social theory. In G.G.A. Bryant, & D. Jary (Eds.), *Gidden's theory of structuration: A critical appreciation*. London: Routledge.

Van de Ven, A. H., Angle, H. L., & Poole, M. S. (1989). *Research on the management of innovation*. New York: Ballinger/Harper & Row.

Varadarajan, P. R. & Menon, A. (1988). Cause related marketing: A coalignment of marketing strategy and corporate philanthropy. *Journal of Marketing, 52*, 58–74.

Von Frantzious, I. (2004). World summit on sustainable development Johannesburg 2002: A critical assessment of the outcomes. *Environmental Politics, 13*, 467–73.

Vogel, D. (2005). *The market for virtue: The potential and limits of corporate social responsibility*. Washington, DC: Brookings Institution Press.

Waddell, S. (2000). Complementary resources: The win-win rationale for partnerships with NGOs. In J. Bendell (Ed.), *Terms of endearment. Business, NGOs and sustainable development*. Sheffield: Greenleaf Publishing.

Waddock, S. (2001). Creating corporate accountability: Foundational principles to make corporate citizenship real. *Journal of Business Ethics, 50*(4), 313–327.

Waddock, S. A. (1988). Building successful partnerships. *Sloan Management Review, Summer*, 17–23.

Waddell, S., & Brown, L. D. (1997). Fostering intersectoral partnering: A Guide to Promoting Cooperation Among Government, Business, and Civil Society Actors. IDR Reports, Vol. 13, 3. Institute for Development Research.

Walsh, J. P., Weber, K., & Margolis, J. D. (2003). Social issues and management: Our lost cause found. *Journal of Management, 29*, 859–881.

Walton, R. E. & Dutton, J. M. (1969). The management of interdepartmental conflict. *Administrative Science Quarterly, 14*, 73–84.

Warhurst, A. (2001). Corporate citizenship and corporate social involvement. Drivers of tri-sector partnerships. *Journal of Corporate Citizenship, Spring*, 57–73.

Warner, M. & Sullivan, R. (2004). *Putting partnerships to work: Strategic alliances for development between government and private sector and civil society*. Sheffield: Greenleaf Publishing.

Wartick, S. L. & Cochran P. L. (1985). The Evolution of the Corporate Social Performance Model. *Academy of Management Review, 10*(4), 758–769.

Webster, M. (2004). *Merriam Webster online dictionary*. Retrieved April 15, 2004, from http://www.m-w.com/dictionary/compatible.

Weick, K. E. (2007). The generative properties of richness. *Academy of Management Journal, 50*, 14–19.

Weick, K. E. & Quinn, R. E. (1999). Organizational change and development. *Annual Review of Psychology, 50*, 361–386.

Weisbrod, B. A. (1998). Institutional form and organisational behavior. In W. W. Powell, & S. Clemens (Eds.), *Private action and the public good* (pp. 69–84). New Haven, CT: Yale University Press.

Westley, F., & Vredenburg, H. (1997). Interorganisational collaboration and the preservation of global biodiversity. *Organisation Science, 8*(4), 381–403.

Whyte, W. F. & Whyte, K. K. (1984). *Learning from the field: A guide from experience*. Beverly Hills: Sage.

Wilson, A. & Charlton, K. (1993). *Making partnerships work*. London: J. Roundtree Foundation.

Wilding, K., Collins, G., Jochum, V., & Wainwright, S. (2004). *The UK voluntary sector almanac 2004*. London: NCVO.

Wilson, A., & Charlton, K. (1997). *Making partnerships work: A practical guide for the public, private, voluntary and community sectors*. New York: The Joseph Rowntree Foundation.

Wood, D. J. (1991). Corporate social performance revisited. *Academy of Management Review, 16*(4), 691–718.

Wood, D. J., & Lodgson, J. M. (2002). Business citizenship: From individuals to organizations. *Business Ethics Quarterly, Ruffin Series* No. 3, 59–94.

World Commission on Environment and Development. (1987). *Our common future*. Oxford: Oxford University Press.

Wragg, D. (1994). *The effective use of sponsorship*. London: Kogan Page.

Wuthnow, R. (1989). *Communities in discourse: Ideology and social structure in the reformation, the enlightenment, and European socialism*. Cambridge, MA: Harvard University Press.

WWF (2004). *Lafarge drops Scottish quarry plan*. Retrieved December 4, 2005, from http://www.wwf.org.uk/news/scotland/n_0000001177.asp

Wymer, W. W., Jr. & Samu, S. (2003). Dimensions of business and nonprofit collaborative relationships. *Journal of Nonprofit and Public Sector Marketing, 11*(1), 3–22.

Yin, R. K. (1994). *Case study research: Design and methods*. London: Sage.

Young, D. R. (1999). Nonprofit management studies in the United States: Current developments and future prospects. *Journal of Public Affairs Education, 5*(1), 13–23.

Zadek, S. (2001). *The civil corporation: The new economy of corporate citizenship*. London/Stirling VA: Earthscan.

Zeng, M. & Chen, X. P. (2003). Achieving cooperation in multiparty alliances: A social dilemma approach to partnership management. *Academy of Management Review, 28*(4), 587–605.

Author Biography

May Seitanidi, Ph.D. (University of Nottingham) is an Assistant Professor at Brunel Business School, London, UK and a Visiting Fellow at the International Centre for Corporate Social Responsibility (ICCSR) at the University of Nottingham. May's background includes researching and working with profit, nonprofit and government organisations. Over the last 24 years her professional and academic interests focus on the interaction between business and nonprofit organisations. May is interested in the application of complexity theory to the study of organisational hybrids, organisational change and the role of functional conflict to achieve sustainable social outcomes across economic sectors. She serves on the editorial review boards of the Journal of Business Ethics and the Journal of Nonprofit and Public Sector Marketing. She was the founder of the Hellenic Sponsorship Centre, publisher of the award winning magazine "Sponsors & Sponsorships", and currently the Editor of the NBP Bulletin, a unique e-publication in the field of partnerships. In 2005 she established an independent group on NPO-BUS Partnerships which numbers over 300 professionals and academics interested in cross sector social interactions (http://tech.groups.yahoo.com/group/NPO-BUSPartnerships/). She lives in London and delights in helping organisations imagine tomorrow's challenges.

Index